Complete Home
Wireless Networking

Windows® XP Edition

About Prentice Hall Professional Technical Reference

With origins reaching back to the industry's first computer science publishing program in the 1960s, and formally launched as its own imprint in 1986, Prentice Hall Professional Technical Reference (PH PTR) has developed into the leading provider of technical books in the world today. Our editors now publish over 200 books annually, authored by leaders in the fields of computing, engineering, and business.

Our roots are firmly planted in the soil that gave rise to the technical revolution. Our bookshelf contains many of the industry's computing and engineering classics: Kernighan and Ritchie's *C Programming Language*, Nemeth's *UNIX System Adminstration Handbook*, Horstmann's *Core Java*, and Johnson's *High-Speed Digital Design*.

PH PTR acknowledges its auspicious beginnings while it looks to the future for inspiration. We continue to evolve and break new ground in publishing by providing today's professionals with tomorrow's solutions.

Complete Home
Wireless Networking
Windows XP Edition

Paul Heltzel

PRENTICE
HALL
PTR

PRENTICE HALL
Professional Technical Reference
Upper Saddle River, New Jersey 07458
www.phptr.com

Library of Congress Cataloging-in-Publication Data

Heltzel, Paul.
  Complete wireless home networking / Paul Heltzel.— Windows XP ed.
      p. cm.
Includes index.
  ISBN 0-13-146153-2
  1. Home computer networks. 2. Wireless LANs. I. Title.
  TK5105.75 .H45 2003
  004.6'8—dc21

                          2003009831

Editorial/production supervision: *Jessica Balch (Pine Tree Composition, Inc.)*
Cover design director: *Jerry Votta*
Art director: *Gail Cocker-Bogusz*
Interior design: *Meg Van Arsdale*
Manufacturing buyer: *Maura Zaldivar*
Publisher: *Bernard Goodwin*
Marketing manager: *Dan DePasquale*
Editorial assistant: *Michelle Vincenti*
Full-service production manager: *Anne R. Garcia*

© 2003 by Pearson Education, Inc.
Publishing as Prentice Hall Professional Technical Reference
Upper Saddle River, NJ 07458

Pearson Education Ltd., *London*
Pearson Education Australia Pty, Limited, *Sydney*
Pearson Education Singapore, Pte. Ltd.
Pearson Education North Asia Ltd., *Hong Kong*
Pearson Education Canada, Ltd., *Toronto*
Pearson Educación de Mexico, S.A. de C.V.
Pearson Education—Japan, *Tokyo*
Pearson Education Malaysia, Pte. Ltd.

Contents

Preface

Setting up a home network used to be a task that only the (very) computer savvy would attempt. How things have changed. There's no reason anyone with multiple computers should go without a network, preferably a wireless one.

Until recently, networking your computers meant stringing something called category 5 Ethernet cable throughout your house. And, in a home situation, few people really needed a network to transfer files from one computer to another. Then along came the Internet, and with it, busy phone lines.

The introduction of broadband, with its very fast digital Internet connections, is helping many home users get more fun and use out of their computers. And, in the home, these broadband connections provide more than enough speed for the whole family to be online at once, *sans* a tied-up phone line. You just need a network to share this speedy Internet access.

It's no wonder that wireless networking is the fastest growing home networking technology today. Wireless networks let you set up an entire local area network (LAN) in the time it takes to drill a hole in your wall. Or hit the stud instead of empty drywall, and then drill *another* hole. Or run out of power on your cordless drill, and take the time to recharge it. Or get frustrated and throw the drill out the window. You get the picture.

With today's wireless networks, you can even take the whole thing with you if you move. Just pack your wireless hardware with all the rest of your computer equipment, and plug it in when you get to your new home or workspace.

One caveat: If your home is very large, you might need to consider that wireless network equipment has a limited range, about 150 feet indoors. A potential workaround for this problem is mixing "no-new-wires" hardware, which uses the existing wiring in your home to network your computers. Generally wireless networking is more convenient than any other technology, but these no-new-wires technologies work very well, and can be a big help. We cover them briefly as well.

So, who is this book for?

- Anyone who could use a simple guide to buying and installing a wireless network
- Those who are setting up their first network of any kind
- People with some computer training but very little network experience
- Folks who have had some experience with a network at the office and would like to see some of the benefits of a network at home

If you're interested in a wireless networking bible, with lots of schematics and lingo, this isn't the book for you. This book will, however, get you up to speed on wireless networks, *quickly*. You'll be up and running in no time.

After reading this book, you should be able to accomplish the following tasks from any room, out by the pool, or sitting in front of the fireplace in your home:

- Surf the web on your laptop
- Print a document on a printer in another room
- Listen to MP3 audio or Web-based video over your network
- Stop burning CDs or using floppies to move files from one computer to another

The premise of this book is to offer *simple explanations of wireless technology* with a focus on installing, setting up, configuring, and troubleshooting your equipment. The emphasis is on a *conversational tone* that points out the best way to avoid problems before you run into them. You'll find lots of *step-by-step instructions* that will help guide you, even if you haven't yet purchased your equipment. Where appropriate, screenshots illustrate where to point, click, and generally wrestle your wireless network into submission.

Again, if this is your first time installing a network, don't worry. You'll find that it's (nearly) painless, and the book will guide you through the troublesome areas. Now let's get to work!

Paul Heltzel

chapter

1

Why Network Wirelessly?

In this chapter...

Wireless networking is one of the fastest growing computer tasks today—and with good reason. Networking your computers with wireless equipment is simple and makes your equipment more useful and portable.

Even if you barely move your equipment across your desk, you'll see the benefits of a wireless network. You no longer have to pull cables or hunt under your desk, usually in an unlit area, trying to find the right ports and cables.

If your office uses a wireless network, creating one at home makes even more sense. Walk in the door, and you are instantly online and able to connect to your other computers and printers—at home or at work.

And while it's true that wireless networking can be slightly more tricky to set up than other types of wired networks, we think you'll find it's worth the extra bit of time it takes. No other networking technology is so flexible and portable

Also exciting is the growth of public wireless access on the road. When you travel, you can connect to wireless networks at hotels, airport lounges, convention centers, and coffee shops, among other locations, and access a high-speed, wireless, connection to the Internet. More and more public access wireless networks are launched each day, and research firm IDC estimates the market will jump from $90 million in 2001 to $60 billion by 2006.

What Is Wireless Networking?

Wireless networking, for the purposes of this book, refers to connecting two or more computers to create a local area network (LAN) using radio transmitter/receivers (sometimes called transceivers). Figure 1.1 shows a wireless network adapter you can connect to a USB port on your computer.

The computers transmit data via radio waves, which allows the sharing of files, printers, and Internet access on every computer in the network. You can move the components of your wireless LAN without losing your connection to the network.

FIGURE 1.1 A wireless network adapter from Linksys.

LANs, Not WANs

It's important that we distinguish your wireless LAN from other types of wireless networks. Wireless wide area networks (WANs) include cell phone networks, so when we talk about wireless networks we're talking about a home network that you can construct in an afternoon, not a cellular voice or data network, which is often called a wireless WAN.

Okay, so we've established the basic components of a wireless home network (LAN): two or more computers connected using wireless network adapters to share files, Internet access, and printers. Wireless networks might also contain other elements, including:

- A scanner, hard drive, or MP3 player shared over the LAN.
- Network hardware that has wires, including a cable modem or a device that connects wired computers together called a hub.

- An access point. This is an important piece of hardware, a wireless bridge that connects a wired network to a wireless network.

In Figure 1.2, we see the components of a simple wireless home network.

Each wireless network device you use creates a cell in which data can be transmitted and received. As you add wireless networking equipment, these cells interlock (Figure 1.3), providing greater distance over which the data is transferred, also called the equipment's *range*.

Later in the book we'll explain the different wireless technologies, called *standards*, and consider their benefits and drawbacks. For the most part we'll concentrate on one technology in this book. The most popular wireless networking technology (or standard for interoperability) is called **Wi-Fi,** or 802.11b. Wi-Fi equipment is great for sharing files, an Internet connection, and peripherals, such as a printer. Wi-Fi equipment is relatively inexpensive (about $80–100 per computer you want to connect).

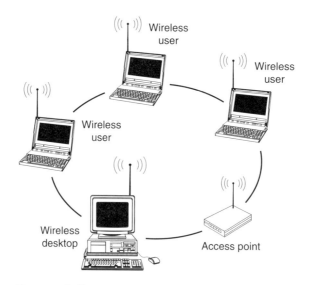

FIGURE 1.2 Computers in a home can connect to each other using wireless network adapters (which include a built-in radio transmitter/receiver, called a transceiver, or sometimes simply a radio).

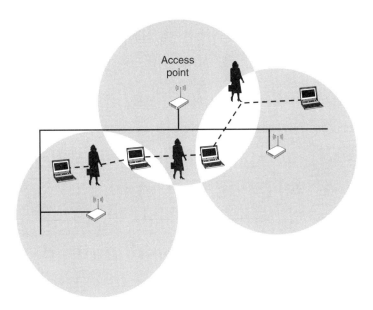

FIGURE 1.3   Users roam around an office, staying connected to their network without being tethered to wired technology.

Benefits of Wireless Networking

Wireless networking offers some benefits that are apparent (no wires) and some that you might not have considered. The most obvious is portability. In particular, if you set up a laptop for wireless networking, you can roam around your house or around a coffee shop. (Starbucks, for instance, plans to offer wireless networks for Internet access for all of its stores in the next few years.)

Moving Your Equipment at Will

It's true that a wireless network will allow you to surf the Internet by your pool, your fireplace, or anywhere else in your house (within about 150 feet, the range of most wireless equipment indoors; more details on that later). More commonly, though, you might find that wireless networking equipment allows you to move your desktop and laptop computers around the house or around your desk or across the room without unplugging cables.

If your office uses a wireless network, you can buy networking equipment for your home that is compatible with your office equipment. That means you can pack up for the day and take your computer home, knowing that when you walk in the door, you will instantly be connected to the Internet, printers, and the other computers in your home that you have connected to your home wireless network. No more plugging and unplugging a laptop from the network.

The convenience that wireless networking provides is obviously a good fit for laptops. But if you want to create a network quickly, and place the equipment wherever you want, without worrying about wires, a wireless network is really the way to go.

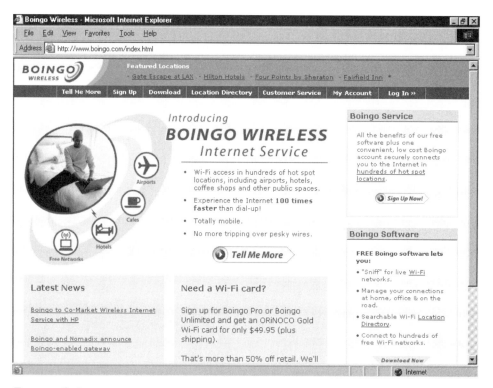

FIGURE 1.4   Boingo offers wireless Internet access to travelers with 802.11b network adapters.

Taking Wireless on the Road

More and more hotels, airport business centers, convention centers, and other businesses are setting up public wireless networks that will help you to access the Internet, at high speed, while you travel. An airline layover can be used to catch up on e-mail and surf the Web. You can turn typical downtime into something really useful, or at least more fun than catching a nap at the terminal.

Companies, including Boingo (Figure 1.4), Sputnik, and NetNearU, are working to create broadband access for travelers. Sometimes called visitor-based networks (VBNs), you can find Internet access when you travel at hotels, convention centers, and airports.

You can use your Web browser to search Boingo (www.boingo. com), for example, to find public wireless access points, or *hot spots*, before you begin your next trip.

Wireless Speed in the Real World

As we discuss various wireless networking technologies, we invariably discuss speed. All the technologies have a maximum speed in which they can transfer data. The data transfer rate is often referred to as bandwidth, or throughput. For instance, Fast Ethernet, a wired technology, can transfer data over cables at up to 100 megabits per second. 802.11a and 802.11g, the fastest current wireless standards, top out at 54 megabits per second. And 802.11b, the most popular standard, transfers data at 11 megabits per second. These speeds are the maximum rate at which data can travel. In the real world, however, wireless networking equipment is likely to offer something on the order of half these speeds.

Compatibility among Manufacturers

As we mentioned earlier, this book concentrates on Wi-Fi (also known as 802.11b) wireless networking equipment. You can purchase 802.11b wireless networking equipment from different manufacturers, and it should work together. I say *should* because a few 802.11b devices are not compatible with the majority of 802.11b equipment. To ensure compatibility among manufacturers look for the Wi-Fi logo, which is displayed on equipment that has met the requirement set up by the Wireless Ethernet Compatibility Alliance (Figure 1.5). You can find out more information about the wireless standard at www.wi-fi.com.

FIGURE 1.5   The Wireless Ethernet Compatibility Alliance is a trade group that promotes 802.11b compatibility.

Importantly, 802.11b does not work with a newer, faster, and more expensive networking standard called **Wi-Fi5** (or 802.11a). Wi-Fi5 is, as you might have guessed, about five times as fast as Wi-Fi equipment. Wi-Fi5 can carry data at a maximum of 54 megabits per second, where Wi-Fi equipment has a maximum speed of 11 megabits per second. Figure 1.6 shows a Wi-Fi5 access point from Proxim. Even newer is 802.11g, which offers the same speeds as 802.11a and is thankfully, compatible with the older and more widely used 802.11b.

As seen in Figure 1.6, Wi-Fi5 (802.11a) equipment, such as this access point from Proxim, works at up to five times the speed of Wi-Fi (802.11b) wireless networking equipment.

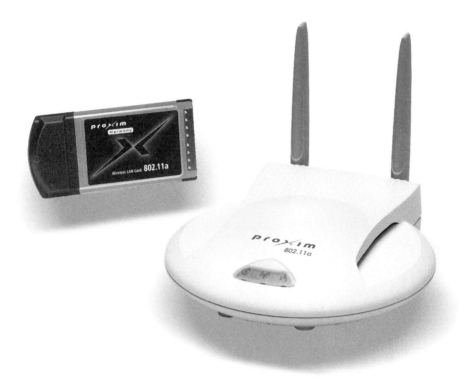

FIGURE 1.6   Wi-Fi5 (802.11a) wireless network equipment.

All of the 802.11 standards offer plenty of speed for surfing the Web, printing documents, and listening to MP3 files. The bottom line is that you should make sure that your products are compatible (by asking the seller or vendor or by looking for the Wi-Fi logo). You'll then be able to purchase equipment from any vendor you choose.

Sharing Your Broadband Modem

The introduction of high-speed Internet access to a wide consumer audience is perhaps the most important new technology for home networks. The term *broadband* refers to fast Internet access technologies, such as cable and Digital Subscriber Line (DSL), which transmit data many times faster than an analog (56-kbps) modem.

Cable modems provide high-speed access, between 300 kilobits to 1.5 megabits per second, over regular TV cable lines. You can see a cable modem, from Linksys, in Figure 1.7. DSL connections provide similar speeds to that of cable over regular phone lines.

Both cable and DSL connections offer fast Internet access that will make your network much more useful. Connecting your wireless network to a broadband connection provides you with a fast link to the Internet without tethering your computer to your modem.

FIGURE 1.7   A cable modem provides much greater speed for connecting to the Internet than a 56-kilobits per second analog modem.

Broadband by Satellite

A less commonly used broadband technology is satellite Internet access, such as that provided by EchoStar (www.echostar.com) and DirecWay (www.direcway.com). Satellite Internet access is often a good option for those in rural areas who cannot get DSL or cable Internet access. Though this is a wireless technology, the companies that sell satellite Internet access provide little in the way of options for sharing Internet access among computers on a network.

Using a home network to share broadband Internet access provides three main benefits:

1. *You won't tie up your phone lines.* Broadband Internet connections do not require that you use your analog phone line to dial up to your Internet service provider.

2. *You get an "always-on" connection.* In most cases, broadband connections are "always-on" so that you don't need to dial in and enter your user name and password each time you want to access the Internet. You just launch your Web browser or e-mail software and get to work.

3. *Multiple computers can share the connection, simultaneously.* Each person in your home or office can browse Web pages, send e-mail, and download files at the same time. You can also set up your network to share an analog, 56-kbps modem, but when sharing Internet access, the faster the connection speed, the better.

These benefits aren't limited to wireless LANs. However, a wireless network connected to a broadband Internet connection offer greater flexibility than a wired network. And since you won't need to wire cable to create your network, you can get started sharing files, Internet access, and peripherals right away.

Sharing Printers and CD and Hard Drives

The capability to share peripherals is another good reason to set up a network. Any printer connected to a computer on your network can be shared by any other computer on the network. In addition, you can use a hardware device, called a printer server, to share a printer on your network without connecting the printer to a PC. Some wireless access points include a port on the back of the device for sharing a printer (see Figure 1.8).

When you set up your network, you can share any drive from any computer on the network. For instance, you can install a software program from a CD-ROM in a desktop computer on a wirelessly connected laptop elsewhere in your house. Or, from your desktop computer, you can access a file stored on the hard drive of a laptop. The files on the hard drive of a networked computer appear in a folder, just as they would on your local hard drive. You can see two shared computers on a network in Figure 1.9. Double-clicking a computer's icon will display all of its shared folders and peripherals, such as printers.

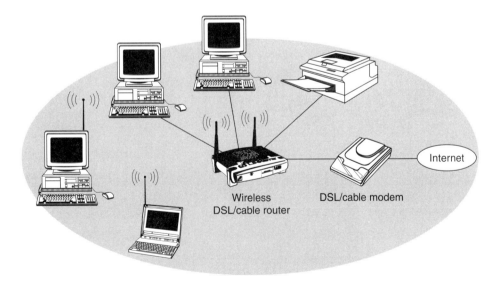

Internet

Wireless
DSL/cable router

DSL/cable modem

FIGURE 1.8   Some wireless access points allow you to connect multiple computers on your network wirelessly and includes a port for connecting (and sharing) a printer on your network.

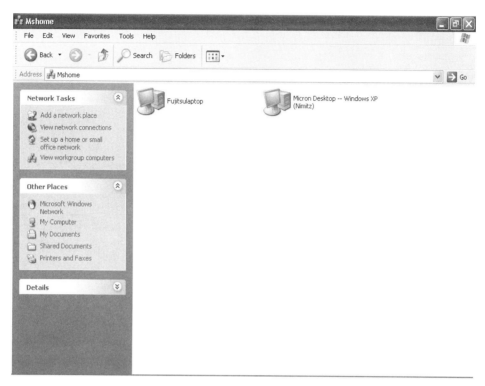

FIGURE 1.9 Accessing files on a computer in your network is as simple as accessing a file on your computer's local hard drive.

You can also share peripherals including scanners, which require special software to use them over a network. UMAX, for example, sells network-ready scanners (not all scanners are made to be used over a network). On the multimedia front, you can purchase network-ready MP3 (audio file) players, such as those from Panasonic and Compaq, which act a bit like a digital jukebox. You can listen to your favorite music, in digital MP3 format, on any computer connected to your network.

After setting up your wireless network, you can generally get more for your money by sharing peripherals rather than buying new ones to connect to each computer. And transmitting data by radio waves beats carrying floppies and CD-ROMs around any day. Setting up a LAN to share files and peripheral access works the same way with a wireless or wired network, and the networking features of Windows make it all quite simple.

Summary

Before you jump into wireless networking, if you haven't already, there are a few questions you might want to ask.

Is your home or office right for a wireless network? In most cases, the answer to this question will be 'yes' (or a qualified yes in a very large home). The range of wireless networking equipment is its greatest drawback. If your home or workspace is quite large, where your wireless networking equipment will be more than 150 feet apart from each other, you will need to consider mixing wired networking hardware and wireless networking hardware. See Chapter 4, which explains how to set up a hybrid network—one with wireless hardware *and* wired hardware (such as Ethernet, phoneline, or powerline networking equipment). Another option: You can add more wireless access points, which will extend the range of your network by another 150 feet indoors (or about twice that outside).

Do you really need wireless equipment? If you simply want to connect two or three desktop computers in the same room, there's not much reason to set up a wireless network. In this case, you'll save money and time with a wired network. However, if just one of those computers is a laptop, and you'd like the ability to work on that computer from any spot in the house, wireless is a great option.

Are you willing to spend extra time setting up wireless equipment? While wireless equipment is more flexible, and portable, than wired networking equipment, it takes more time to set up. That said, once your wireless equipment is up and running, you won't need to spend much time maintaining it. If you are interested in the flexibility of wireless, it's well worth the extra time for setup.

Even if you have never set up any kind of networking equipment, don't be intimidated. You'll quickly see how simple it is to connect your computers and begin sharing data wirelessly. In the next chapter, we look at how wireless networking works, get some jargon out of the way, and start planning a wireless LAN.

chapter

2

Getting Ready

In this chapter...

- ✔ Learning Some Basic Networking Terms
- ✔ How Wireless Networking Works
- ✔ File Sharing
- ✔ Internet Sharing
- ✔ Networking Securely

To make sure your wireless networking goes smoothly, you should learn a few networking basics. For the most part, you don't need deep technical knowledge of how each software and hardware component does its job. Yet there are a few new terms and concepts that you should understand.

A few technical definitions will come in handy as you shop for equipment, and when you need to call your manufacturer's technical support for assistance. Of course, by keeping this book around while you set up your equipment, you can keep those tech support calls to a minimum.

More importantly, you're about to get a sense of how a simple wireless network works. We'll consider how wireless networks transfer data, and how you can share files, Internet access, and peripherals. We'll also look at how wireless networking security works. By the end of this chapter, you should have a pretty good understanding of what you need to know about wireless networks. Later in the book, we delve into more detail, and more how-to advice on setting up and maintaining your equipment.

Learning Some Basic Networking Terms

Here are some terms we use throughout the book. Don't worry too much about remembering all of these now. We'll reintroduce them in context. If nothing else, you can use them to intimidate the salespeople at your local computer store.

LAN (local area network): A network of computers in one location, usually a home or office.

Network adapter: Also called a network adapter card (Figure 2.1) or network interface card (NIC), this is a device you use to connect a computer to a network.

Access point: This hardware device allows wireless network cards to connect to a wired network (Figure 2.2). An access point has a wired component (an Ethernet port) and a wireless component (a radio that allows wireless network adapters to connect to the network).

FIGURE 2.1   A wireless network adapter plugs into your computer and transmits and receives data by radio waves.

Router: A router is a hardware device, or a software program, which allows one network to connect to another. In a home network, you can use a router to connect your LAN to the large network of interconnecting networks called the Internet. You can buy an access point with a built-in router (Figure 2.3). Your router will allow you to share a single Internet connection among all the computers connected to your network.

FIGURE 2.2   An access point acts as a wireless go-between, connecting a wireless network to a wired network.

FIGURE 2.3   This access point includes a built-in router.

Gateway: A gateway can be hardware or software that allows multiple computers to access the network. In most cases, on a LAN, your gateway is a router. Your gateway could also be a single computer sharing its Internet connection with the other computers on the network.

Protocol: A protocol is a language used by a network to send and receive data. TCP/IP (Transmission Control Protocol/ Internet Protocol) is the protocol used to transfer data over the Internet. You can also use TCP/IP as the protocol for your home network, for sharing Internet access, files, and printers.

Broadband modem: Unlike a 56-kilobit per second modem, which sends and receives data over analog phone lines, a broadband modem sends a digital signal over your telephone or TV cable wiring, depending on whether you use a DSL or cable modem, respectively.

Ethernet: A wired standard for networking hardware. Why mention a wired technology? Some of your equipment, such as a broadband modem, will connect to your access point by an Ethernet cable (see Figure 2.4). Until recently, Ethernet was pretty much the only technology available for networking your computers.

Hub: A wired hardware device that is used to connect multiple computers to your network. A hub has ports, usually four or more, in which you connect an Ethernet cable, which looks like a phone cable but is slightly thicker (Figure 2.5). You can use a

FIGURE 2.4 Ethernet cabling is often used to connect networking hardware, including a broadband modem, hub, or an access point.

FIGURE 2.5 A four-port hub.

hub to connect wired devices, such as a cable modem or router, to elements of a wireless network, such as an access point. An access point works a bit like a wireless hub and connects a wireless network to a wired network.

Now that you have a basic understanding of these network components, you will be able to choose the best network for your home or office, make smart equipment purchases, and set up your network with minimum headaches.

How Wireless Networking Works

The wireless network adapter you use to connect a PC to your wireless network transmits data by radio waves. But unlike an FM radio transmitter, your wireless networking equipment sends a signal that can only be picked up for about 300 feet if there are no obstructions (and this kind of performance is usually achieved outside only).

As with wired networking technology, such as Ethernet, data is transmitted over a wireless network in pieces, called data packets. Each network adapter has its own unique serial number, called a *MAC* (media access control) address. You can see the MAC address of your wireless network adapter, as it's usually printed on the underside of the adapter. The data packet contains the data being sent as well as the address of the sender and recipient.

Wi-Fi (802.11b) wireless networks, as well as 802.11g equipment, communicate over the unlicensed 2.4-GHz radio band. They share the band with other home electronics, including cordless phones and microwaves.

Wi-Fi5 (or 802.11a) wireless networks use the less crowded 5GHz band, and therefore are less affected by home electronics. While operating at up to five times faster than Wi-Fi (802.11b) equipment, the hardware looks just the same (Figure 2.6).

Until newer technologies such as Bluetooth (see sidebar) become more prevalent, wireless networking will mean setting up a Wi-Fi system in most home applications.

Other Wireless Networks

Many PDAs, cell phones, and laptops can transmit data at very high frequencies using beams of infrared light. Devices that use an infrared wireless technology called IrDA (Infrared Data Association) can communicate with each other. You could print a document from a laptop to a printer, for instance, or transmit (beam) contact information from one PDA to another. IrDA is a line-of-sight technology, which means that the devices much be lined up in a straight line to communicate.

Bluetooth is another technology used to transfer data wirelessly. Bluetooth is found in PDAs and cell phones and can be used to transfer data to laptop and desktop computers and printers. Like 802.11b and HomeRF networking equipment, Bluetooth operates in the 2.4 GHz band, does not require a line of sight between components, and can pass through walls.

Figure 2.6   An 802.11a wireless network adapter for a laptop.

Bluetooth technology is sometimes referred to as a personal area network (PAN). The technology has a more limited range (about 10 meters) and less speed (720 kilobits per second) than wireless LAN equipment, such as 802.11b.

Wireless network adapters can communicate directly with each other. Each network adapter acts as a transmitter and receiver, and data is broadcast in a cell. As the cells interlock, the network's range expands. A wireless network in which the network adapters communicate directly with each other is called a *peer-to-peer network,* or is said to be working in *ad-hoc* mode. Wireless networks can also work in *infrastructure* mode, which requires the use of a hardware device that communicates with each network adapter, called an access point (Figure 2.7).

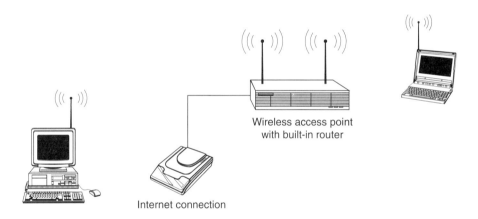

Wireless access point with built-in router

Internet connection

FIGURE 2.7   A simple wireless network.

Once your wireless networking equipment is set up, you can share data and peripherals just as you would over a wired network.

File Sharing

By enabling file sharing on your PCs, you can view and copy shared files from any computer on your network. Windows operating systems all offer File and Printer Sharing service. All you need to do is enable it.

You can share folders (and therefore the files inside them) or entire CD-ROM drives or hard drives. However, sharing an entire hard drive is not typically recommended. Select the folders you want to share, rather than giving access to your entire hard drive and all your system files.

If you do quite a bit of file sharing, you'll want the fastest networking equipment available, currently 802.11a or 802.11g.

Since wireless connections are less secure than wired networks, you will need to take special considerations to keep someone from accessing your network. Before you begin sharing files on your wireless network, you should read Chapter 12 to see how to protect your data and your privacy.

Internet Sharing

Internet sharing is one of the most useful—and most fun—ways of using your wireless network. Here's how it works.

Your Internet Service Provider (ISP) provides you with an *IP address*, a unique number that identifies your computer on the Internet. An IP address is four sets of numbers, from 0 to 255, separated by periods, such as 26.0.162.255. The problem is that your ISP likely only provides you with *one* IP address. So how can more than one computer use the same IP address?

Typically, the solution is to use a *router*. Each computer on your network will connect through a software or hardware router rather than connecting directly to the Internet. The router provides an IP address for each computer on your network. To the rest of the world, it appears as if you're using just one IP address.

You can run a software router on one of your computers that will hand out an IP address to each computer. A more convenient option is to buy a hardware router.

Using a hardware router will allow you to share Internet access without leaving a computer on to do the job. However, in a wireless network, using an access point with a built-in router is an even more convenient option. It can often save you setup time and money, too.

Networking Securely

Maintaining security over a network usually takes more than one form. You can mix some of the following means to keep unwanted users off your network. We explain, step by step, how to use these security measures in Chapter 12. In the meantime, here's a quick overview of ways to protect your network.

Wireless Encryption: Networking hardware comes with software to encode data over the network so that it can't be read by an unintended recipient. The data is scrambled at the source, then descrambled by the recipient. The technology standard for wireless encryption is called WEP (Wired Equivalent Privacy). You can enable wireless encryption on your network using the software that comes with your wireless network adapter (Figure 2.8). You'll see how to enable WEP on your network adapters, and, if you have one, your access point, in Chapter 12.

Setting the SSID: The SSID (Service Set Indentifier) on a wireless network is a name that identifies your network. To access the network, the SSID on each computer has to be the same (Figure 2.9)

Filters (such as MAC address): Each wireless network adapter (as well as wired network adapters) come with a number that uniquely identifies them. You can set up your network so that only computers with MAC addresses you enter are allowed on the network (Figure 2.10).

Firewalls: A firewall on your network puts a layer of protection between you and a hacker. Firewalls can be software you install on a computer, or they can be built into a router, or used as a standalone firewall hardware device. Whether you decide to install software or hardware, a firewall will help keep outsiders from accessing your network.

Antivirus software: You'll want to use an antivirus program on your individual computers to keep viruses from spreading on your network. A virus can wreak havoc by invading your computer's hard drive and, in some cases, sending out copies of

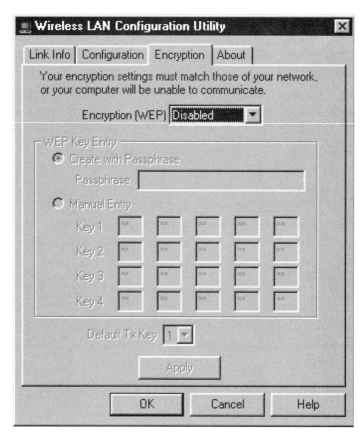

FIGURE 2.8   Enabling wireless encryption using the software that comes with a wireless network adapter.

itself via e-mail. A good antivirus program will catch a virus when you receive it, and keep it from spreading.

Sadly, none of these measures is foolproof. Wireless networking equipment is inherently more insecure than wired networking equipment. Why? Wireless networks don't require a physical connection to your network to be hacked. A smart hacker with the right know-how can tap into your data as it is transferred over the airwaves. Remember,

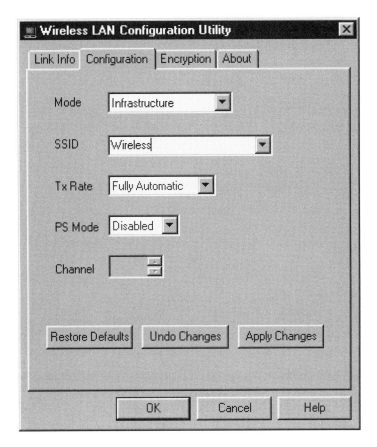

FIGURE 2.9   Each wireless network adapter needs to provide the same SSID to be allowed access to the network.

anytime you connect your computer to a network, there is a certain level of risk involved.

That said, just because wireless networks are easier to break into doesn't mean you shouldn't use them. It just means you should be aware of the risks involved and take appropriate measures. You want to, at the very least, keep your neighbor (or neighboring offices) from accidentally accessing your wireless network. In Chapter 12, you'll find out how to make your wireless network difficult to hack.

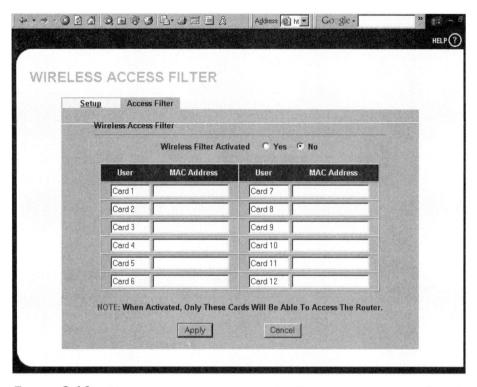

FIGURE 2.10 You can set your router to only allow access to a network card with a MAC address that you enter.

Summary

This overview of wireless networks should help you understand what you'll need to start a wireless network. You can use an access point, router, and network adapters to equip your home or small office and be up and running in a day. Once your equipment is installed, you can begin sharing files and Internet access. Of course, you should consider using a firewall and encryption to keep your data safe from hackers— or anyone else in range of your wireless network. In the next chapter, we'll consider how range varies depending on the environment where you install your wireless LAN.

chapter

3

Wireless Considerations

In this chapter...

Wireless networking offers some obvious convenience over its wired counterparts, such as Ethernet networks and those that work over the telephone or via electrical wiring in your house. However, wireless networking also has some drawbacks. It's a good idea to know a little about the pros and cons before you start your network.

S ince wireless equipment works over radio waves, the signal sent can only travel over a limited distance. Unlike the powerful radio transmitters that broadcast the stations you listen to in your home and car, a wireless network radio can only broadcast a signal over a short distance indoors (and about twice as far outside, if there are fewer obstructions than inside). Again, when we speak of wireless networking generally, we're talking about Wi-Fi (802.11b) equipment, the most widely used wireless standard today. The more obstacles that get in the way—walls, brick, concrete, steel, and to a lesser extent, glass—the less distance your wireless network will be able to cover.

The software that comes with your wireless network adapter will show you the strength of your signal. Your software might look slightly different than the figures you see in this book, but they all work in a similar way: When you get farther away from another wireless network adapter or access point, the software will display decreased signal strength, and your data transfer rate will drop (see Figure 3.1).

The area over which a wireless network can broadcast, sometimes called its *range,* is a factor you need to consider when setting up your network. Your network's limited signal range can be a problem in a very large house.

You can work around this range in several ways. For example, you can add an access point to extend the range of your network. You can also mix wired and wireless networking hardware to expand the distance over which you can use your network. We'll discuss these options in this chapter.

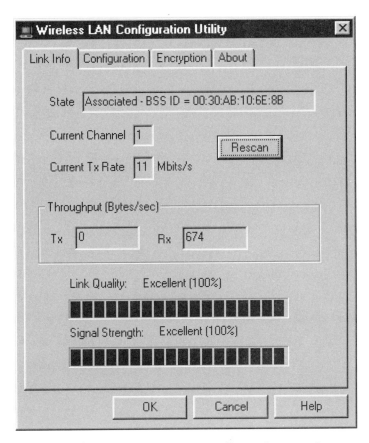

FIGURE 3.1 Here you see the software for a wireless network adapter in close proximity to the access point. The signal strength shows 100 percent (or "excellent").

Your Wireless Equipment's Range

Let's say that you have a wireless network adapter plugged into your laptop computer's PC Card port. As you walk around your home, increasing the distance between your wireless networking equipment, the speed that the network adapter can transfer data drops incrementally. This speed decrease is by design, allowing the wireless network adapter to maintain a reliable network connection, albeit slower than if the adapters were closer together (see Figure 3.2).

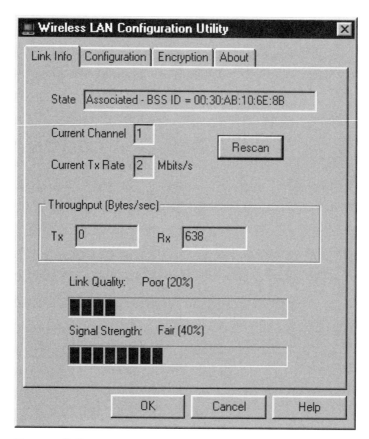

FIGURE 3.2   Your network's radio signal drops off as distance is increased. The indicator shows a transfer rate (labeled here as *Tx rate*) of just 2 megabits per second. Compare that to 11 megabits per second when the signal is at full strength.

The speed drops off at set rates, from a top speed of 11 megabits per second to 5.5, 2, and 1 megabit per second. If you move beyond 150 feet (or less depending on the number of obstructions in your home or office) the signal is too weak and the network connection is lost.

Ad-Hoc vs. Infrastructure

Wireless networks typically work in one of two configurations (sometimes called *topologies*): **ad-hoc** or **infrastructure.** The topology or *mode* you choose depends on whether you want your PCs to communicate directly or with an access point.

In ad-hoc mode (Figure 3.3) data in the network is transferred to and from wireless network adapters connected to PCs. An ad-hoc network is also called a peer-to-peer network. Here are some of the benefits of an ad-hoc network:

- Ad-hoc networks are simple to set up. Plug in your wireless network adapters, configure the software, and you're off and running.
- Ad-hoc networks are inexpensive. You save the cost of purchasing an access point.
- Ad-hoc networks are fast. Throughput rates between two wireless network adapters are twice as fast as when you use an access point.

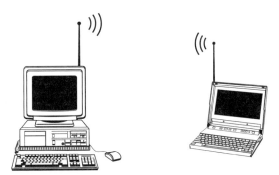

FIGURE 3.3   An ad-hoc network with two computers communicating directly.

Now that we've considered the benefits of ad-hoc networks, let's consider the road most people will follow when creating a wireless network. You can increase the range of your wireless network by adding an access point. Wireless networks that use an access point are in *infrastructure* mode (Figure 3.4).

An infrastructure network enables you to:

- *Connect to a wired network.* An access point (Figure 3.5) lets you easily expand a wired network with wireless capability. Your wired and wirelessly networked computers can communicate

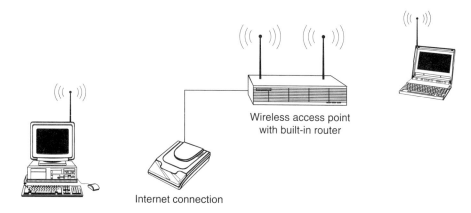

Wireless access point
with built-in router

Internet connection

FIGURE 3.4   More common than an ad-hoc network, an infrastructure
network includes an access point.

with each other. This is the most obvious strength of an infrastructure setup.

- *Extend your wireless network's range.* Placing an access point in between two wireless network adapters doubles their range.

- *Utilize roaming ability.* If you add multiple access points to your network, as you might in an office or large home, users can *roam* between interlocking access point cells, without ever losing a connection to the network.

- *Share the Internet.* Probably the most useful device in a simple wireless LAN is an access point with a built-in router and firewall. The router allows you to share Internet access between all your computers, and the firewall hides your network, helping to keep network-savvy hoodlums at bay. Some of these multifunction access points include a (wired) hub as well, for plugging in several computers connected to your network by Ethernet. You can purchase a device like this for less than $200, a real steal for all the use you get out of it.

What about drawbacks? An infrastructure network takes a bit more work than setting up an ad-hoc network. Infrastructure net-

FIGURE 3.5   This access point from D-Link
can be used to connect a cable or DSL modem
to provide Internet access to the wirelessly
connected computers on your network.

works cut the data transfer rate about in half, because of the time it
takes to send the signal to and from the access point rather than directly to its destination, as in an ad-hoc network. The other drawback is
expense: Infrastructure networks are more expensive than ad-hoc networks because you have to purchase an access point. As we already
mentioned, however, that expense is well worth it for all the benefits
an access point provides.

How Walls, Concrete,
and Steel Affect Your Signal

The matter of how far your network adapter's radio signal will travel
varies greatly depending on the layout of your home or office. Here are
a few considerations to keep in mind while you design your wireless
LAN.

The signal of wireless network adapters does not require that the network adapters have a direct line of sight between them. The signal bounces off objects and can pass through walls. That said, it is important to keep in mind that:

- Concrete and steel will seriously degrade the signal.

- Glass, to a lesser extent, will also weaken the signal.

- Books and other solid objects often found on a desk, if covering the adapter, will degrade the signal significantly.

All wireless network adapters come with software that will tell you the strength of the signal, often with a visual representation on your desktop (see Figure 3.6). As you move your equipment around your

FIGURE 3.6   Network adapters often come with software to measure the strength of your wireless signal. At the bottom right of the screen, an icon shows the wireless network adapter is receiving a signal.

home or office you can easily see how strong your signal is. The signal is shown as a percentage, and often also is represented by a color, with green showing an "excellent" signal, yellow being a "fair" signal, and a red indicator showing no link to the network. We talk more about working with this software in Chapter 4.

Now we'll consider how the speed your software displays translates to what you'll actually get.

Maximum Speeds vs. Real World Speeds

When you shop for wireless networking hardware you will typically see the equipment's top speed listed on the web site or on the packaging at the store. You should realize that the top speed is rarely, if ever, achieved.

It's not uncommon, in an average-sized home, to see 802.11b equipment, with a top speed of 11 megabits per second, averaging 3–4 megabits per second. That's still likely faster than your cable or DSL modem Internet connection. In a home environment this is plenty of bandwidth for typical applications like surfing the Internet, printing, and copying MP3 files from one computer to another.

In my own home, I've found that 802.11a equipment (with a top speed of 54 megabits per second) often operates at less than half that speed. With the addition of an access point, the speed halves again because of the roundtrip communication between network adapter and access point (see Figure 3.7).

The data sent over your network includes more than just the files you are downloading from the Internet, or copying from one computer to another. Data also needs to be transferred that includes information on how and where the data is supposed to arrive, as well as confirmation that the data has reached its destination. This extra data, along with potential interference from other wireless devices, and network latency make it difficult to reach the theoretical maximum speeds of wireless connections.

FIGURE 3.7   When you add an access point, the network adapter's radio sends and receives data from the access point.

Getting the Most Out of Your Wireless Signal

In your own home, there's no guarantee of what sort of bandwidth you'll see on your wireless network, due to distance and obstructions (concrete in the basement, steel beams, your dog). You'll want to position your equipment to get the strongest possible signal. Here are a few ways you can maximize and extend the strength of your signal:

- Where possible, place equipment in a direct line of sight with as few obstructions as possible. Wi-Fi equipment does not require a line of sight, but works faster when placed in a direct line between radios.

- Place your access point as close to the center of your network as possible (Figure 3.8).

- Install additional, wired equipment where convenient. See Chapter 4 for information on hybrid networks (Figure 3.9).

- Add additional access points to extend your network's range.

If you consider yourself a bit of a do-it-yourselfer, you can purchase larger antennas to improve signal strength. You can mount some

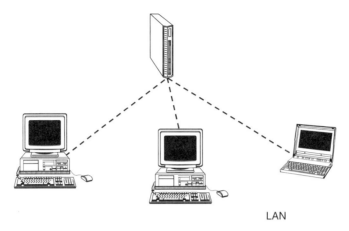

LAN

FIGURE 3.8 Center your access point in your network for the best possible signal.

of these antennas on your ceiling to provide a less-obstructed signal, although the installation is beyond the focus of this book. Keep in mind too that you can purchase another access point for about the same price (about $200) as a larger antenna.

FIGURE 3.9 You can buy no-new-wires equipment that uses the existing wiring in your house, such as phoneline network adapters, to extend the range of your network. Here we see a phoneline network adapter from Netgear.

802.11a, 802.11b, or 802.11g?

Most of the information here holds true for the newer, faster Wi-Fi5 (802.11a) equipment as well. However, while 802.11a offers faster maximum speed than 802.11b (54 megabits per second rather than 11), 802.11a equipment has a shorter range, about 60 feet.

The shorter range of 802.11a means you'll need to purchase more access points than you would with an 802.11b network. That's a significant concern, considering 802.11a equipment is a newer technology, and more pricey than older 802.11b hardware (see Table 3.1). That price difference won't mean much to companies that need the fastest wireless networking equipment available, but it will likely keep most home users buying 802.11b equipment.

TABLE 3.1   Features of 802.11b, 802.11a, and 802.11g

	Top Speed	Approximate Range	Pros	Cons
802.11b	11 mbps	150 feet	Good for simple Internet, file, and printer sharing	Less speedy than 802.11a or 802.11g
802.11a	54 mbps	60 feet	Band means less interference from cordless phones and Bluetooth devices that use the 2.4GHz band	More expensive than 802.11b hardware
802.11g	54 mbps	About 10 percent less than 802.11b	Fast & compatible with widely adopted 802.11b	Newer technology is always more expensive. Can suffer from electronic devices that use the 2.4 GHz band, such as cell phones

Most home users will benefit from the relatively low cost of 802.11b wireless networking hardware, and its speed should suffice for home Web surfing, file sharing, and printing. Office users who need to connect laptops for roaming around the office might be more interested in 802.11a or 802.11g equipment for its speed in copying large files, such as databases, photographic images, and video files.

Summary

Before creating your network, you should consider the limiting factors of wireless equipment. Those in vary large homes or workspaces will need to consider adding additional equipment, such as wired networking hardware or extra wireless access points, to provide a strong signal, and therefore faster link, to their wireless networking hardware. Those attracted to the faster speeds of new 802.11a and 802.11g equipment will do well to consider its shorter range and higher expense, compared to older 802.11b wireless networking equipment, which is commonly found both in home and office networks.

Wireless Hardware and Software Setup

In this chapter...

Here's where things get fun. Seriously. You get to start connecting the wireless hardware you've been reading about in the previous chapters. Whether you have just two computers you want to network or an office full of PCs that need configuring, now you can find out how to get your computers communicating with each other.

Most of the work involved in getting your wireless network adapters communicating happens in this chapter. You'll also find tips on buying equipment that will make setup easier. An access point with a built-in router is an example.

We'll also consider how to install your equipment and look for connection problems. In the following chapters we'll work through changing settings on your PCs to share data, but the connection building starts now.

Connection Options

You have three basic choices when choosing a wireless network adapter. Depending on the wireless standard and type of computer you are using (see Table 4.1) your choice will to some degree be made for you. The options are a PC Card, USB, or PCI Card. Let's look at these different types of network adapters in detail.

TABLE 4.1 Choosing Wi-Fi Network Adapters

If you have...	For use with...	Then use...
802.11b	Laptop	PC Card
802.11b	Desktop	USB
802.11a or 802.11g	Laptop	PC Card
802.11a or 802.11g	Desktop	PCI Card

PC Card

PC cards (Figure 4.1) slide into one of two PC card slots you'll find on a laptop. Note that the PC card slot is sometimes called a PCMCIA slot (PCMCIA stands for Personal Computer Memory Card International Association). PC card wireless adapters (Figure 4.2) are

Figure 4.1   A PC card wireless network adapter from 3Com.

FIGURE 4.2   A wireless network adapter that connects to the USB port on your computer. Photo courtesy of Netgear.

great for notebooks because they are small and easy to install. When you hit the road, you can leave the card in your laptop's PC card slot. (If the laptop won't fit in your bag with the network adapter plugged in, you can pop it out for storing in the pocket of your laptop case.) You can also use a USB wireless network adapter with a laptop, though USB adapters tend to be larger and include a cord you probably don't want.

USB

For connecting desktops to a 802.11b network, USB is a very good choice. USB (Universal Serial Bus) network adapters are inexpensive and easy to install. PCs that run Windows 98 or later (when you purchased them) will typically have two USB ports. The ports are hot-swappable, meaning you can plug in equipment and unplug it without rebooting the machine. That said, you will sometimes need to restart your computer for some network software to recognize the network adapter. There is one caveat to using USB with wireless networks: You can't use it for 802.11a or 802.11g networks, because USB isn't fast enough to carry data for these wireless networks, which transfer data at 54 megabits per second. The maximum throughput of USB is 12 megabits per second.

PCI Card

PCI (Peripheral Component Interconnect) cards are sometimes used to connect wireless network adapters to desktop computers. You can see a wireless PCI card network adapter in Figure 4.3.

Connecting a PCI card to a computer is slightly trickier than using a USB network adapter. You must open the computer case and install the card in an open PCI slot inside the computer. It's not terribly complicated, but it takes more time, especially compared to installing a USB network adapter. Early 802.11b PCI network adapters sometimes required that you install a PCI adapter in your computer, then plug in a PC card network adapter into the PC card adapter. Remember to check out Table 4.1 for a quick look at what equipment to buy for different computer setups.

FIGURE 4.3   A PCI Card wireless network adapter.

Setting Up Hardware

Now we'll look at the basic steps for installing your wireless networking hardware. Keep in mind that the equipment you purchase might dictate that you vary the order of these steps, so follow the instructions that come with your networking equipment. This, however, will give you a quick idea of how the process works.

1. Plug in your network adapter.

2. Install the drivers for your network adapter.

3. Reboot your computer.

Your hardware vendor might recommend that you install software for your network adapter before installing the network adapter, or the other way around. It's important to follow the instructions that come with your card. Your equipment might not work otherwise.

Software Overview

You won't need any special software to begin using your wireless network equipment. Using the software that comes with your wireless network adapter, you can quickly establish a link between two adpaters (in ad-hoc mode) or connect to the access point (in infrastructure mode).

Again, follow the directions that come with your equipment, but here is the basic idea. There are three settings that are most important. If these are not set correctly and uniformly on each of your network adapters (and your access point, if you have one) your network will not work. These are Mode, SSID, and WEP.

Open the configuration software that comes with your hardware (in fact, it might be running when you start your PC, as many programs are set by default to launch when Windows starts).

Choose Infrastructure or Ad-Hoc (Figure 4.4). As described earlier, this tells the wireless network equipment whether you are connect-

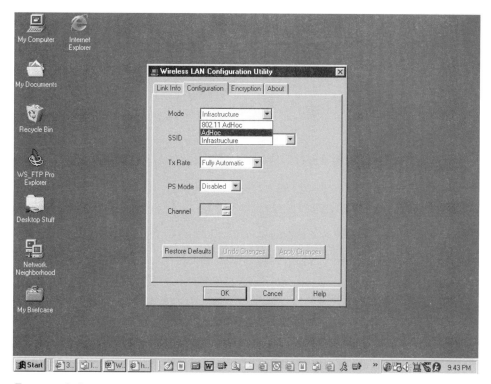

FIGURE 4.4   Choose Infrastructure or Ad-hoc mode.

ing directly to network adapters (ad-hoc mode) or through an access point (infrastructure).

The SSID (Service Set Identifier) is the name of your network (Figure 4.5). You must use the same SSID on all your equipment or your hardware will not be able to connect to the network. The SSID can be discovered quite easily by a hacker who can use software to "sniff" the SSID over the airwaves. So it doesn't offer protection from unauthorized users within range of your network.

Enable or disable WEP (Figure 4.6). You need to make sure your wireless encryption settings are the same for all your equipment. Your network adapters and access point all need to have WEP on or off, otherwise your equipment won't link up to each other.

With these settings uniform, your computers should be ready to communicate over the network. You might need to change some network settings on your Windows computers to begin sharing files, which we discuss in some detail in Chapters 7 and 8. But the hard work of software configuration is complete. Congratulations!

FIGURE 4.5   Set the SSID. If you choose "ANY" the network adapter can connect to any available 802.11b network.

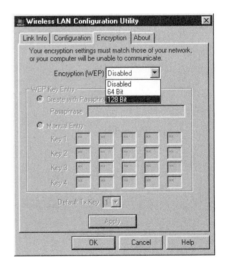

FIGURE 4.6   Enable encryption.

Using Ad-hoc Mode

Setting up an ad-hoc network can be a quick and inexpensive way to get your network started. Here are the steps necessary for setting up your hardware in an ad-hoc network.

1. Plug in the network card.

2. Install drivers for the network card.

3. Reboot the machine.

4. Set configuration software to ad-hoc mode.

After installing one wireless network card, you would follow these steps for each computer that you want to connect. How many computers can you connect? Ad-hoc mode has a theoretical maximum of 256 nodes (computers connected to the network).

> **TIP:** Wireless networking cards can eat up your laptop batteries quickly. If you are disconnected from your wireless network, as when traveling, and working off batteries, consider disconnecting your wireless network adapter.

Keep in mind that if you want to share Internet access between computers in ad-hoc mode, you will need to use connection sharing software, such as the Internet Connection Sharing (ICS) software that comes in Windows 98 Second Edition and later (including Windows Me, 2000, and XP). Also, one of your computers will need to stay on to share the connection.

Working with Access Points

An access point is a great way to extend your wireless network, either to connect to an existing wired network or to share a broadband modem. Installing an access point should be as simple as plugging the power adapter into an outlet and connecting an Ethernet cable to a computer or hub. The trickier aspects of access points come when you start working with them. So it's good to become familiar with a few potential hangups before you get started.

There are two basic types of access points:

1. An access point (Figure 4.7) used as a bridge between a wireless network and a wired network.

FIGURE 4.7   An access point from 3Com.

2. An access point with a built-in router. The router lets you share Internet access to all your computers (Figure 4.8).

Once you change any settings as instructed in the installation procedure that comes with your equipment, such as setting encryption or the SSID, you should be ready to go.

FIGURE 4.8   A Linksys access point with a router.

TIP: Not all access points are capable of communicating with each other. You can purchase an access point that will communicate with another access point (sometimes called Ethernet over AP or AP-to-AP). This is helpful if you wish to set up multiple access points in your home or office, to extend your wireless network's reach and do not plan to connect each access point to a wired network. If this is important to you, ask whether your access point offers this feature before you buy.

You can purchase an access point that has a router and a built-in hub for connecting wired computers to the router. In the next section, we look at how this setup might benefit you.

Access Points with Built-In Routers

Unless you already own a router, you should seriously consider purchasing an access point with a built-in router. An access point alone can't share Internet access, so a built-in router will make your life considerably easier for just a bit more money than an access point alone.

If you choose to install an access point, you can always plug it into a wired (Ethernet) router (see Figure 4.9), which is quite inexpensive, but will add one more connection and more bulk to your setup.

FIGURE 4.9 You can connect a wired (Ethernet) router, like this one from Linksys, to an access point by Ethernet cable.

It should also be mentioned that routers are sometimes called *gateways*. A gateway is a broader term that refers to hardware or software that provides Internet access to the computers on your network.

Most routers, once plugged into your network, are configured using your Web browser. You simply enter an address, often 192.168.0.1.

Your router should have a feature called NAT (Network Address Translation). NAT allows you to share one IP address provided to you by your Internet Service Provider (ISP). NAT creates a series of subaddresses for use on your network. These IP addresses

(192.168.0.1–192.168.0.253) do not appear on the Internet. This is handy for two reasons:

1. Your computers are hidden from hackers who might want to access the computers on your system.

2. You can make use of one IP address to put as many computers on the Internet as you want. To your ISP, it appears that just one computer is using the Internet. You don't have to pay for multiple IP addresses.

So once you plug in your router, you can open up your Web browser and configure the device (Figure 4.10). You might be instructed by your ISP to adjust settings, such as entering your user name and password so that you won't have to on each machine you connect. Your ISP, or

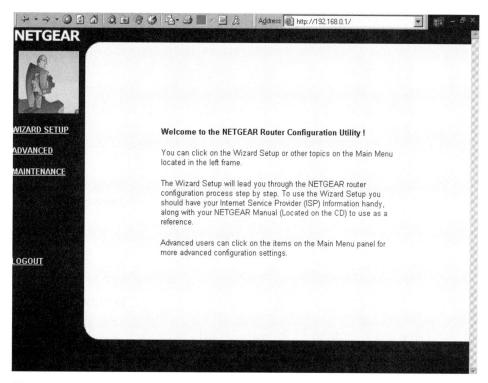

FIGURE 4.10   Configuring a router by connecting through a Web browser.

your hardware manufacturer, will need to provide you with specific instructions. In many instances your router will be set to connect, using the default settings, right when you plug it in.

Here are the basic steps for setting up a router.

1. Open your Web browser.

2.. Enter the IP address of your router.

3. Follow the directions from your ISP. You might be told to enter in an IP address, or more likely, you'll use the default setting, which is to allow the ISP to assign an IP address.

Later in the book we'll discuss some other features that are likely offered by your router, such as the capability to allow only network adapters you choose to access the network (and keep anyone else from joining it).

> **NOTE:** Some ISPs, especially some cable Internet access providers, will want to charge you for each computer that connects to the Internet. Find out before you sign up for broadband Internet access whether your ISP will allow you to use a router. If your ISP prohibits use of a router to share your ISP address, avoid that provider. For instance, if the local cable company won't allow you to use a router, check out the price for DSL access.

Troubleshooting

Once all your equipment is connected you might find that one or more network adapters is not connected. Again, you will typically see a green light in your system tray, or some similar indicator, that shows your computer's wireless network adapter is connecting with the access point in infrastructure mode or with other network adapters in ad-hoc mode.

1. Check your connections.

2. Make sure all equipment is set to ad-hoc mode if there is no access point.

3. Make sure your cable or DSL modem is working and connecting to your ISP. Most modems have three lights, and they should all be lit.

Placing Your Equipment

It is important to discuss where to place your equipment, and give you a visual idea of what sort of equipment you can connect to extend the range—and the usefulness—of your wireless network.

Here we consider how to place equipment for simple two-PC wireless networks, where the computers communicate directly together, as well as multi-PC wireless networks that use both wired and wireless hardware to put a networked computer in every room in the house. We'll look at a handful of scenarios that you might run across, and then read recommendations on how to set up equipment in those situations. Also, we'll look at the costs for each setup so you can see what you're getting into.

Creating a Peer-to-Peer Network

Scenario: You have just two computers and don't need to share an Internet connection (Figure 4.11). Or, you plan to share the Internet connection on one of your computers using Internet connection sharing (ICS).

A peer-to-peer network has the dual benefit of being the least expensive and the easiest to set up. You just need two wireless network adapters ($30–$120 for 802.11b network adapters; slightly more for 802.11a or 802.11g).

The main problem with a peer-to-peer network is that its range is limited to the distance over which each network adapter can broad-

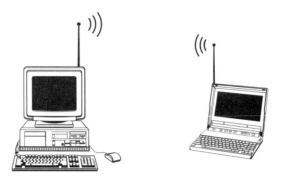

FIGURE 4.11   A peer-to-peer network is
easy to set up and inexpensive.

cast. To extend your range, you'll want to set up an infrastructure network, which we look at next.

Creating an Infrastructure Network

Scenario: You want to share files and printers between more than two
computers and would also like to share an Internet connection using a
broadband modem (Figure 4.12). An infrastructure network will probably work best in this situation.

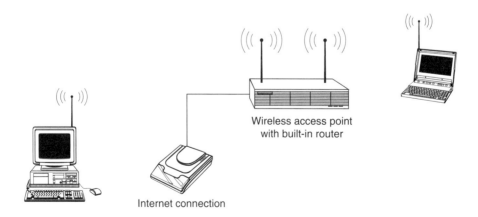

Wireless access point
with built-in router

Internet connection

FIGURE 4.12   An infrastructure network lets you add more computers and is
useful for sharing a broadband modem.

Infrastructure networks take a bit more configuration than setting up an ad-hoc network, but not much more setup time. Also, the additional benefits of an infrastructure network—primarily that an access point allows you to connect to wired devices, such as a cable modem, or wired networks, such as an office Ethernet network—certainly make them worth the time to set up.

Adding an 802.11b access point will cost you between $60–$250, plus the cost of the network adapters you will need for each computer.

An infrastructure network can be made even more useful by incorporating Ethernet, powerline, or phoneline networking equipment.

Setting Up a Hybrid Network

Scenario: You have desktops that don't really need the portable capability of wireless networking hardware and laptops that could benefit from wireless network adapters. You could mix no-new-wires technologies, such as phoneline and powerline, which use your existing house wiring. These technologies are available at your local electronics or office supply store. You can connect powerline and phoneline network equipment with your wireless access point and network adapters (Figure 4.13).

Phoneline network adapters cost less than $50 per adapter. Powerline adapters are a bit more expensive than wireless ones, at around $120–$150. Ethernet adapters are the budget choice (as well as being the fastest technology mentioned here), costing less than $20 per adapter.

Hybrid networks are probably the most difficult to set up. However, they are potentially the only kind of network that will get the job done in some environments, such as very large homes that are already wired for phone and electricity but are too large for a wireless network to cover. Another instance where a hybrid network can be useful is an office setting that has an existing wired network.

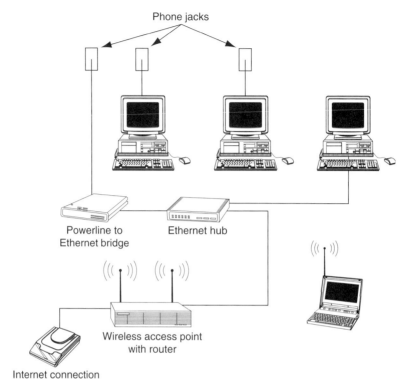

FIGURE 4.13   A hybrid network lets you combine wired, or no-new-wires technology, and wireless network hardware.

Connecting Network Multimedia Devices

Scenario: You want to add a standalone device, or two, to your wireless network. These devices might include an MP3 player, such as the Compaq IPaq Music Center for about $800, or a printer server, which can run from $75 to $200.

Not all of your equipment needs to be connected to a PC. You can use devices that will speed up your network, such as a wireless printer server, which needs no physical connection to the network. You can also make a device that everyone uses, such as a scanner or MP3 player, more easy to access by connecting it to an access point that has ports for a hub (many do, and it won't add much to the price).

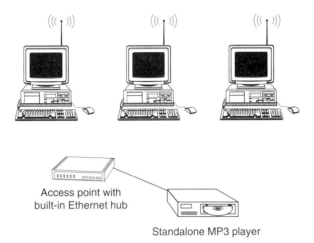

Access point with
built-in Ethernet hub

Standalone MP3 player

FIGURE 4.14 You can connect standalone elements, such as a network MP3 player, using a hub with a built-in Ethernet hub.

In the next section we'll consider how older equipment can be used with equipment that is introduced later, and typically helps speed up elements of your network.

Mixing 802.11 Equipment

Scenario: You have an investment in your existing Wi-Fi equipment when an attractive new wireless technology hits the market. Thankfully, you don't need to trash your existing hardware. You can likely integrate the equipment to help you save money and get more life out of your initial networking hardware.

If you add an Ethernet hub to your network, you can connect more than one access point to the hub, which can serve as a bridge between 802.11b, 802.11a, and Ethernet networks. Thankfully, new network technologies, such as a faster form of 802.11b networking hardware that operates at 22 megabits per second, is starting to arrive, and the technology is backward-compatible with older 11 megabits per second

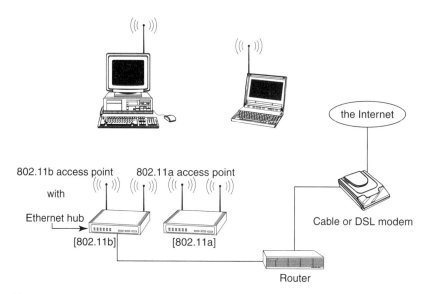

FIGURE 4.15   Mixing wireless technologies in the same network might require the use of a hub.

802.11b equipment. And, as mentioned, 802.11b and newer, faster 802.11g networking hardware are compatible.

In the future we will see new wireless technologies that might or might not be able to work with your older equipment. But be assured that you can buy useful, fast wireless equipment later, and you can add a hub or buy backward-compatible hardware to meet your needs.

Summary

What kind of network will you build? It depends on how many computers you want to connect and what you'll do with them. Ad-hoc networks let you connect just two computers. Infrastructure networks expand your network with an access point. Hybrid networks let you mix wired and wireless technologies. And networks with standalone devices help make your network more fun or useful without connecting the hardware to a PC.

After choosing and plugging in your network adapters (and installing the accompanying software) you can connect your network adapters to each other or to your access point. Choose ad-hoc or infrastructure mode using the software that comes with your network adapters, and your wireless network will be up and running.

5

Making Use
of Wireless Standards

In this chapter...

It's time to choose the wireless networking technology that is best for your home or office LAN. You'll find most of these technologies stacked side by side at your local computer store, so it's a good idea to know how they differ from one another. And we'll look at new wireless standards that increase speed and compatibility.

While we tend to favor the 802.11b standard for its price and widespread use, we'll discuss other wireless networking standards, including 802.11a and 802.11g. Each has benefits and drawbacks that you should know before you buy your equipment.

We'll also look at infrared, a ubiquitous but little-used technology for transferring files and printers. This older technology's speed pales in comparison to that of the latest wireless networking standards, but you can find it on many laptops and PDAs.

A Good Start: 802.11b

Inexpensive and reliable, 802.11b can help you get started with wireless networking. It's efficient (faster than your Internet connection) and easy to use.

However, 802.11b isn't the only wireless networking technology available. Other 802.11 wireless networking technologies are becoming more common. And you can use infrared light to connect laptops and PDAs or to print from a laptop.

The 802.11b standard isn't going anywhere soon, and its widespread adoption will likely complement the rest of the technologies explained in this chapter.

Next we consider using 802.11b to get your PDA connected to your network.

Wirelessly Networking a PDA

So, now that you're working with a wireless network, why not consider cutting cables with your PDA as well? You can toss your USB or (much) more awkward serial cable connection and connect your PDA's address book, e-mail, and appointments wirelessly to your PC.

Despite a famous commercial's promise of faxing and sending e-mail from the beach, most PDAs do not wirelessly connect to the Internet. For that you'll need to purchase a wireless modem (or connect a serial cable to your PDA and cell phone).

All PDAs running the Palm or PocketPC operating systems have the ability to *synch* data. That means you can click a button on a stand called a *cradle* (or select an option using software) to replicate the data on your PC to your PDA and vice versa. That way, all your data is current no matter whether it's entered on your desktop or your PDA.

Most PDAs have some wireless capability for transferring data over short distances, but here's the catch (and it's a pretty big one). Your handheld might use an infrared standard that is different from the one found on most laptop computers. And, most desktops do not have infrared capability. That wireless capability on your PDA is mostly good for transferring data between PDAs since two PDAs of the same type will use the same infrared standard.

So, how do you network your PDA wirelessly? Here's where 802.11b comes in handy again. You can buy an 802.11b add-in card for your PDA for about $90 (Figure 5.1).

FIGURE 5.1   You can use a compact flash network card to add your PDA to your wireless network.

For example, you can purchase an adapter for your PDA from Netgear, D-Link, SMC, and others. Once you do, follow these three steps to connect to your wireless network:

1. Install the drivers for your network adapter.

2. Plug in the wireless network adapter in your PDA.

3. Start your PDA and follow the instructions for your network adapter to connect to your network.

As soon as you walk in your wireless network's range, you'll be ready to go. In addition to synching up wirelessly, some PDAs will allow you to surf the Web and get e-mail (capabilities vary widely by model). You can download Web pages and e-mail to view later, when you don't have an Internet connection. All in all, 802.11b can be a handy and elegant solution for getting more use out of your PDA.

HomeRF

HomeRF, a competing wireless standard to 802.11b, initially showed quite a bit of promise due to its multimedia capabilities. The companies that produced HomeRF equipment, however, recently decided to stop making these products.

The wireless networking standard can be used with cordless phones, since it shares the 2.4GHz band used by some cordless phones. In addition, HomeRF (Figure 5.2) is quite handy at playing multimedia, such as MP3 audio files. The networking standard gives priority to multimedia over other network communication, which helps to offer smooth playback.

So, why didn't everybody use HomeRF? A fair question. The bottom line is that 802.11b was already widely used in corporate settings, which gave it a built-in audience at the office and offered double-duty at home. But, most importantly, 802.11b offers greater speeds.

The first version of HomeRF worked at 1 megabit per second. Not bad for Web surfing, but too slow for file sharing. The first version of

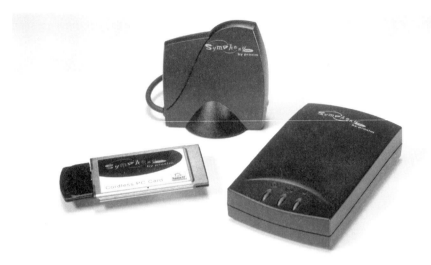

FIGURE 5.2   A HomeRF access point from Proxim.

802.11 (the precursor to 802.11b) offered speeds of 2 megabits per second. The second version of 802.11b (running at 11 megabits per second) hit the ground well before HomeRF 2.0 (10 megabits per second) was released. The delay in releasing the second version of HomeRF gave 802.11b a considerable speed advantage, and consumers took advantage of it.

Moving Up: 802.11a

A constant complaint about wireless networking has long been that it's too slow, but no longer. 802.11a equipment is now widely available, and it offers a maximum speed about half as fast as FastEthernet (100 megabits per second).

802.11a equipment is especially good for corporate users who need a speedy technology to extend their existing wired networks by adding 802.11a access points (Figure 5.3). The price for 802.11a is nearly twice that of 802.11b, so most home users won't likely consider it. In addition, at the time of this writing, only laptop adapters were available, furthering the point that the technology is first being marketed to corporate

FIGURE 5.3   You can add wireless
capability to an existing wired network
using an 802.11a access point with a built-
in hub.

users who want to use fast wireless technology for laptops, not desktops.

Like 802.11b equipment, you can use an 802.11a access point to
combine wired and wireless hardware. For instance, you could use an
802.11a access point to share a cable modem.

You could also connect the access point to a hub, which opens up
quite a few options. For example you could connect to the hub phoneline or powerline networking hardware, creating a mixed wireless and
wired network that will easily offer access to the network in every
room in your home.

Next, we check out the upcoming 802.11 standards that are meant
to both speed up and simplify wireless networking.

Newcomer: 802.11g

As mentioned earlier, the 802.11b and 802.11a wireless networking
standards are incompatible. Here are a couple of potential issues you
should keep in mind when considering 802.11b and 802.11a equipment.

If you already have an 802.11b network, upgrading to 802.11a might
mean replacing your existing wireless network adapters to the newer

standard. Or you can maintain two wireless networks operating at two different speeds.

The two standards have different ranges. Despite being faster, 802.11a has a shorter range (about 60 feet) and speeds drop off faster than 802.11b as you move equipment away from each other.

A new 802.11 technology helps unify two network standards and make them work better. Here's how.

The proposed 802.11g, like 802.11a, offers a maximum speed of 54 megabits per second. However, unlike 802.11a, 802.11g operates in the 2.4 GHz frequency band, and is compatible with 802.11b.

In Table 5.1 you'll see some of the other networking standards you can choose as well as their benefits and drawbacks.

TABLE 5.1   Wireless Networking Standards

Standard	Speed	Pros	Cons
HomeRF	10 mbps	Multimedia capabilities	Not as fast as 802.11b, no longer made
802.11b	11 mbps	Ubiquity/ Inexpensive	Slower than other wireless technologies
802.11a	54 mbps	Speed	Incompatible with 802.11b
802.11g	54 mbps	Compatible with 802.11b; greater range than 802.11a	Slightly more expensive than the other 802.11 technologies

There are alternatives to RF (radio frequency) technologies such as 802.11 if you want to simply transfer files between two PCs. Infrared technology uses a beam of light to send data between computers. In the next section we'll look at how to quickly connect two computers to trade files using infrared.

Sending Files via Infrared

Because infrared (IR) is a line-of-site technology, your equipment needs to be positioned with the infrared transceivers pointing directly at each other. And infrared won't go through walls and floors the way radio frequency technologies like 802.11 will.

To send data between two computers you need: a built-in infrared port, commonly found on most laptops, or an add-on infrared port, such as those sold by Extended Systems (see Figure 5.4).

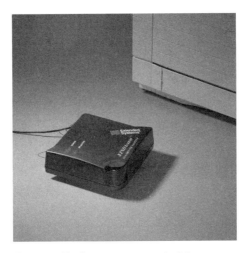

FIGURE 5.4   An Extended Systems infrared transceiver for printing wirelessly (www.extendedsystems.com).

TIP: Note that the standard called IrDA (Infrared Data Association) is used by many laptops and PDAs. Some Palm OS PDAs use another infrared technology, however, so check into the model you're considering if infrared data transfer and printing is important to you.

Here's how to transfer data between computers using Windows Direct Cable Connection software that's built into Windows 98. First you need to enable the Infrared capability on a laptop or other device with an infrared transceiver. Here's how:

1. From the Control Panel (Start, Settings, Control Panel), double-click Infrared.

2. Click the Options tab and choose Enable infrared communication. Remember the ports shown at the top of the dialog box on which infrared will be enabled (Figure 5.5). You'll need this information later.

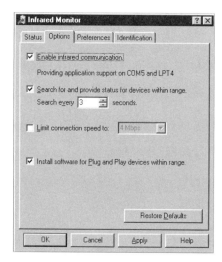

FIGURE 5.5   Enable Infrared capability.

3. Choose Start, Programs, Accessories, Communications.

4. Choose Direct Cable Connection. If you don't see this applet, you might need to install it from your Windows operating system disk. (Open the Windows Control Panel and open the Add or Remove Programs applet. Click Windows Setup tab, and then choose Communications. Click the Details button and select Direct Cable Connection.)

5. The Direct Cable Connection wizard appears (Figure 5.6).

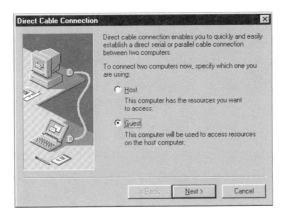

FIGURE 5.6   Start the Direct Cable Connection wizard.

You'll be asked what port you want to use. Choose the port that Windows enabled in Step 1. The wizard will then step you through creating a wireless infrared connection to the other PC.

In Windows XP, infrared capability is enabled by default on a PC with an infrared transceiver. Align your laptop, digital camera, or PDA so that the two devices are within a meter of each other, with their red windows facing each other. When they are close enough to communicate, an icon appears in the taskbar that shows the devices can communicate.

You can trade data between a PDA and laptop using infrared, but the interfaces are so unwieldy (and compared to 802.11b, so much slower) that you will be much better off to use (802.11b).

Printing through Infrared

Printing wirelessly could mean one less cord to carry around and connect, no small matter if you often travel with your laptop. To print using infrared, you need:

- An infrared port on your computer.
- A built-in infrared port on your printer.

Hewlett Packard sells some inkjet printers that include infrared ports. However, since most printers do not have them, you can also buy an add-on infrared port for your computer, such as those sold by Extended Systems, for about $40.

Summary

Although 802.11b is the most widely used wireless standard today, other standards exist that can help you connect wirelessly. 802.11a presents a faster way to go wireless, but is incompatible with 802.11b. Infrared technology can be used for connecting computers and printing, but it is much slower than 802.11b and does not pass data through walls. Newer, faster technologies, such as 802.11g, work with older technology (802.11b) and will help reduce incompatibility problems.

6

Finding Hardware and Help

In this chapter...

✔ Conducting Your Research

✔ Finding Reliable Reviews

✔ Buying Equipment Online

✔ Ensuring Interoperability

✔ Finding Help on the Internet

✔ Checking Out the Web

✔ Finding Answers through Newsgroups

✔ Getting Assistance from Mailing Lists

✔ Looking Toward the Future

No matter how much research and shopping you do, one fact will remain: technology changes quickly. You might feel like your equipment is obsolete between the time you walk out of the store and the time you open the packaging to install it. To some degree, it can't be helped. The latest, fastest, and (hopefully) most user-friendly products are constantly in development, and they always have the annoying habit of being cheaper than the equipment you bought last month. Don't worry too much about obsolescence. Do your research, and then jump in.

Once you know what kind of wireless hardware you require, you can start researching the specific models you want and then find them at a reasonable cost. The Web can help.

If you buy smartly, you can also better your chances for interoperability down the line. Knowing that wireless technologies such as 802.11b and 802.11a are incompatible could save you a costly mistake. Especially when buying online, make sure that the product you are purchasing is what you think it is. Price looks too good to be true? Make sure you're getting the latest version of the technology. More than one penny-pinching buyer has found their wireless networking equipment uses older and slower technology that might not be worth the price break. Be a smart consumer, and make a phone call to hardware makers and retailers to get the facts.

Conducting Your Research

Since technology changes quickly, your first stop should be a Web site that specializes in networking. You'll be able to find out what the experts think will best suit you, find the latest news, and often find reviews of wireless networking hardware.

The following sites will give you a good, up-to-date view of the current wireless networking offerings from vendors. Start your search here and you'll find more about wireless network setup and troubleshooting that you probably will ever need. Especially if you're just starting out, you should get to know these sites.

HomeNetHelp (www.homenethelp.com)

HomeNetHelp offers excellent tutorials and troubleshooting (Figure 6.1). The site does a particularly nice job of mixing its own advice with feedback from users, who offer insight on the unexpected hangups you

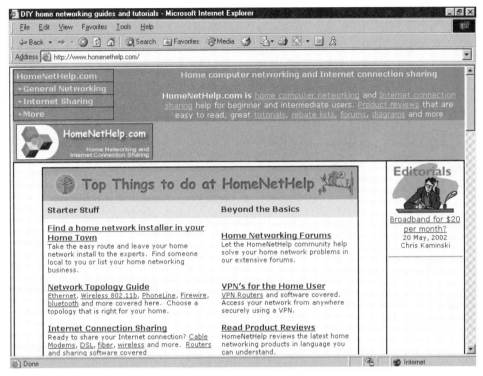

FIGURE 6.1 Stuck? Check out HomeNetHelp.

might find when installing hardware. HomeNetHelp is a great place to find what might be called the secret handshake—that one small hardware tip or software setting that, when you know about it, can bring your network to life. Without the secret handshake, you're often left in the dark.

About.com (www.compnetworking.about.com)

About.com can point you to a number of helpful, and often-updated, web sites for all kinds of networking, both wireless and wired. This networking resource weeds out the lackluster sites and helps you find ones that will help you get started—or get you out of a jam (Figure 6.2).

FIGURE 6.2   About.com has become an invaluable source of good advice—and good links—on the Web.

homePCnetwork.com (www.homepcnetwork.com)

Focused primarily on wired home networking, homePCnetwork.com can still help you find out smart ways to get your computers talking, including connecting PCs and Macs.

Practically Networked (www.practicallynetworked.com)

Practically Networked is a must-see site when you're getting your network off the ground. Simple explanations and diagrams make this site one of the best around for planning, troubleshooting, and maintaining your first wireless network.

Finding Reliable Reviews

Your best bet for finding wireless networking equipment is to read comparative reviews. You'll find a number of independent news sources offering their buying advice free on the Web. Here are a handful of computer trade magazines that put wireless networking equipment through lab testing and real-world applications to help you know what you'll be getting before you purchase.

CNET (www.cnet.com)

A great starting place. CNET's Web-only approach might lack some of the depth of a longer article in a print magazine, but easily makes up for it with frequently updated content (Figure 6.3). CNET offers coverage of both PCs and Macs, as well as, to a lesser extent, the Linux operating system.

FIGURE 6.3   CNET offers wide-ranging advice and wireless hardware reviews.

PC World (www.pcworld.com)

Favoring concise, plainly spoken advice and tips over jargon, PC World is a great place to find news about new wireless networking hardware, reviews of the latest equipment, and how to set it up. You'll also find the full issue of the print magazine on the web site.

PC Magazine (www.pcmag.com)

PC Magazine has some of the best-known reporters and columnists in the computer industry. Strong, thorough coverage of the latest technologies. Highly recommended.

FIGURE 6.4   TechWeb caters to information technology professionals.

TechWeb (www.techweb.com)

TechWeb tends to offer more of a hobbyist or information technology (IT) approach to its news, review, and how-to articles. That said, news coverage is top notch, and a great place to learn about the latest trends in wireless hardware and software (Figure 6.4).

Once you've finished researching hardware, it's time to get your equipment. Whether you're just starting out or upgrading equipment, the sites in the next section can make the job easier.

Buying Equipment Online

Let's assume you've found the hardware you need and have a general idea what it will cost you from reading over the trades on the Web. Now it's time to find a fair price and purchase your hardware. A good place to start is with an online comparative pricing service.

CNET Hardware

Formerly called Computers.com, this site was created and is maintained by CNET. The site mixes reviews (rating hardware on a 10-point scale) with links to its shopping site, Shopper.com. A nice mix of editorial and buying advice.

Shopper.com

Unlike Computers.com, Shopper.com (Figure 6.5) focuses on price comparisons among online sellers rather than equipment reviews. The site does offer user reviews of sellers, however, which can help you find reputable vendors.

PriceWatch (www.pricewatch.com)

This underused site offers a no-nonsense design, fast loading pages, and an efficient search engine. Most importantly, PriceWatch displays

FIGURE 6.5   Shopper.com points you to the latest prices.

some of the lowest prices you'll find from sellers. A great site when you're ready to buy (Figure 6.6).

Once you have a range of prices in mind, look at some of the larger online stores, such as Outpost or Buy.com or PC and Mac Connection (you can access both through www.pcconnection.com), and consider their special offers. See if a free shipping deal or an attractive rebate might help reduce your bottom line.

Finally, once you have prices in hand, check with your local electronics retailers, such as Circuit City, Best Buy, and office stores, such as Staples or OfficeMax. They might have prices similar to those you find online. And consider that paying a slightly higher price at your local office store might be worth it, especially if you need to return a non-functioning wireless network adapter, which happens.

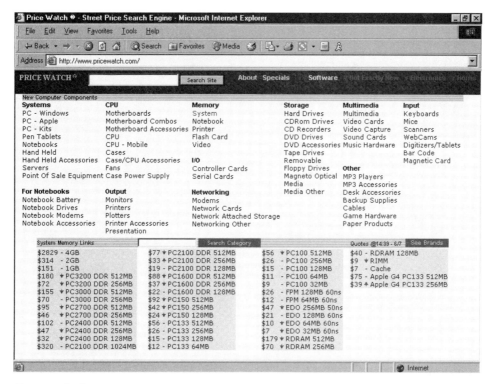

FIGURE 6.6   PriceWatch can help you determine a fair price for the wireless equipment you want.

How Much Is This Going to Cost?

What should you expect to spend? As a general rule, newer technologies will always be more expensive. So, if you're looking to adopt the latest, fastest wireless technology, you'll pay for it. That said, as more people adopt wireless technology, prices will fall. Networking with 802.11b is a bargain, and will become even more economical as new wireless products are introduced.

Table 6.1 presents a quick look at what you can expect to spend on wireless networking hardware.

TABLE 6.1   *Wireless Equipment Costs*

Equipment	Approximate price range
802.11b network adapters	$30–$120
802.11b access points	$60–$250
802.11b access points/routers	$80–$300
802.11a network adapters	$60–$250
802.11a access points	$200–$350
802.11a access points/routers	$120–$400
802.11g network adapters	$80–$140
802.11g access points	$120–$400
802.11g access points/routers	$140–$450

Prices and technologies change quickly, of course. Table 6.1 should give you a general idea of what you'll be looking to spend, but check the comparison shopping sites mentioned in this chapter to find the best deals on hardware.

Ensuring Interoperability

One of the most worrisome aspects of purchasing equipment is that it won't work with newer equipment, essentially negating your investment. There's no use trying to predict what new technology will be available two years from now, but you can work to ensure the ability to upgrade your existing equipment.

Thankfully, all wireless access points include Ethernet ports. Since Ethernet is the most common and least expensive means to network computers, you can mix or *bridge* wireless and Ethernet network hardware.

For instance, you can purchase an Ethernet-to-phoneline or Ethernet-to-powerline bridge. These hardware devices will allow you to use "no-new-wires" technologies to expand your network. Phoneline equipment is less expensive than newer powerline equipment, but either can be a big help if you have available outlets or phone jacks. We discuss networks that use more than one technology, called hybrid networks, in more detail in Chapter 4.

You might find that you'll get more use, and longer life, out of your wireless equipment if you use its capability to connect to Ethernet and bridge other networking technologies.

Finding Help on the Internet

There's an obvious benefit to the easy access and timeliness of wireless networking information on the Internet, and now that you don't have to wait to get online anymore, you should make use of it.

The 802.11 wireless networking standards are in flux, and new information comes out about them every day. New equipment, some that will bridge different wireless standards, is introduced at a fast pace as well. And you'll want to stay on top of pricing as costs come down and you look to purchase equipment that will allow you to expand your network.

The Web isn't the only place to look for wireless networking information. Resources such as online newsgroups often offer specific information about your make and model of equipment. Mailing lists are also a big help and give you access to savvy network users that can help you when you have a specific question and are in a hurry.

Checking Out the Web

Need an updated driver or a firmware upgrade to your router? Check out your vendor's web site and while you're on the Web, compare vendors' online tech support. Odds are you'll be using a network hardware vendor's support site to download drivers and find help for your equipment. Make sure the vendors you buy from offer lots of guidance and easy-to-find software downloads.

Here is a list of common wireless network equipment makers.

3Com (www.support.3com.com)
Compaq (www.compaq.com/support/files/networking)
D-Link (www.dlink.com/tech) (Figure 6.7)

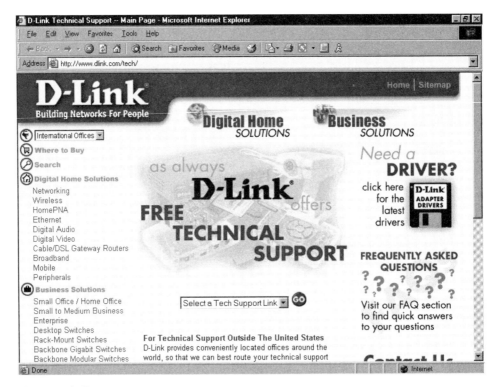

FIGURE 6.7   D-Link's Web site.

Intel (www.support.intel.com)
Linksys (www.linksys.com/download)
Netgear (www.support1.netgear.com/netgear1) (Figure 6.8)
Proxim (www.proxim.com/support/software)

Industry Groups

If you need to find information on a particular type of wireless network technology, an industry trade group web site can often help. Need to find out if your equipment is interoperable? These sites can guide you.

WECA (www.weca.net)

The Wireless Ethernet Compatibility Alliance concerns itself with making sure 802.11 products can interoperate. Products that pass muster get the "Wi-Fi" logo. So, for example, if you buy a Wi-Fi-

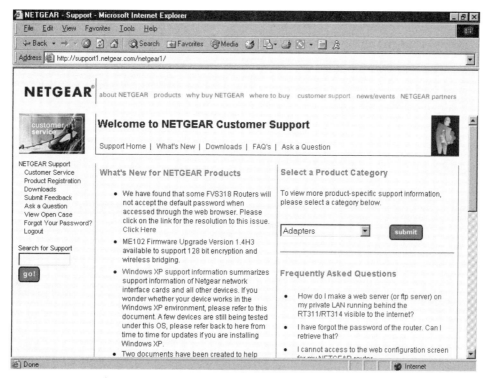

FIGURE 6.8   Netgear's Web site.

certified 3Com router and a D-Link Wi-Fi network adapter, they should work together right out of the box.

Bluetooth (www.bluetooth.com)

The official Bluetooth website can provide you with the latest on this personal wireless networking technology. Bluetooth is useful for sharing data between phones, PDAs, printers, and other electronic devices.

Finding Answers through Newsgroups

When your network equipment isn't working or you have questions before you purchase hardware, newsgroups are a good place to check in. You don't need to take the time to configure a newsgroup reader. You can read through and search newsgroups using your Web browser.

Google Groups (www.groups.google.com)

Google Groups (Figure 6.9) offers a searchable interface to news-groups, with an archive that goes back to the mid-1990s. Since you're probably more interested in getting the latest information, the site can help there too, offering the capability to sort by date. Google is particularly helpful when you have a question that relates to a specific make and model of equipment. In many instances, you'll find that your question has already been answered.

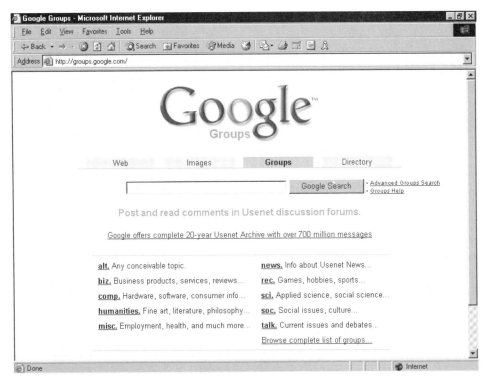

FIGURE 6.9   Google Group's Web site.

comp.std.wireless

The comp.std.wireless newsgroup offers plenty of advice on all sorts of wireless networking subjects. Subject matter tends to be a bit more technical than most of the other resources mentioned here. If you are interested in seeing whether your question has already been answered in the newsgroups, check out the Google Groups search (previous listing).

Getting Assistance from Mailing Lists

If you don't mind a few more messages in your inbox every day, a mailing list can be a good way to keep informed about wireless LANs. You can typically sign up to receive the e-mail messages as they are submitted, or all at once, in digest form.

Publicly Accessible Mailing Lists (www.paml.net)

Not sure where to start? You can find mailing lists here, as well as an archive of messages. The Mail Archive Site (Figure 6.10) offers a searchable archive of mailing lists.

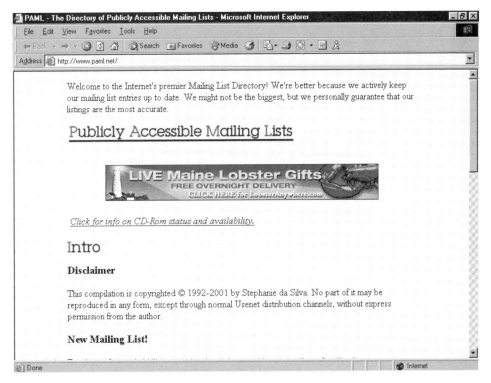

FIGURE 6.10 Publicly Accessible Mailing Lists' Web site.

Wireless LAN (WLAN) Mailing List (www.groups.yahoo.com/group/wirelesslan)

The Wireless LAN Mailing List is a good place to sign up for news and user concerns about wireless networks. Discussion includes topics such as Wi-Fi, Bluetooth personal networking technology, and connecting PDAs to wireless networks. Messages are archived by date at the site. You'll need a (free) Yahoo account to join this group.

Looking Toward the Future

If you're looking for the latest trends in wireless equipment, including newer, faster technologies that are coming soon, check out these sites.

FIGURE 6.11   802.11- Planet's Web site.

80211-Planet.com (www.80211-planet.com)

What better place to start than a web site that has 802.11 in its name? A recent discussion focused on 802.11b products that will get a speed boost in future editions and will be able to communicate with older and slower 802.11b hardware? Is it worth the extra money? Check out the site's hardware reviews (Figure 6.11) to find out.

IDG.net (www.idg.net)

Staying on top of wireless LAN technology can be daunting. You can keep abreast of new technologies that are introduced nearly daily at IDG.net, a web site that pulls together the content of a number of trade magazines and is fully searchable.

These resources should help you get more work done with your wireless LAN and spend less time troubleshooting. Good surfing to you.

Summary

Buying the right equipment at a reasonable cost is easy if you know where to look. First stop, the Internet. Here you'll find great advice from experts and users on which wireless technologies will work best for your environment. Once you know what technology you want, it's time to search for the make and model you'll purchase. With this information in hand, you can find a great deal by using web sites that consolidate pricing information from different vendors. Final tip: Keep an eye out for coupons (at sites such as www.dailyedeals.com) that can offer discounts, rebates, and free shipping.

A range of free, Internet-based web sites and services can help you stay on top of your wireless LAN. Both professional sources, such as trade magazines, and user-based resources, such as mailing lists, are available and can inform you of new equipment you should know about, judge its effectiveness, and show you how to best put wireless network hardware to good use.

Step-by-Step
Networking
with Windows 98

In this chapter...

You say you've got Windows 98 computers? Well, you're in luck, because in this chapter, we'll walk you through the standard setup for connecting a Windows 98 computer on a wireless network.

Windows 98 computers are quite easy to network. In addition, if you use Windows 95 or Windows Me, you should be able to follow these directions to set up those operating systems on your network. More recent versions of Windows, such as XP and 2000, are set by default to appear on the network. We look at setting up those operating systems for wireless networking in the next chapter.

There is an important difference between Windows 95/98 and Windows Me. The more recently introduced Me offers the Home Networking Wizard, which can handle some of this operation for you. In addition, Windows Me has "Network Places," a folder that shows your available network resources, whereas Windows 95 and 98 have "Network Neighborhood."

You'll want to have your operating system CD-ROM handy, as you might be prompted after making network settings to copy files from the CD to your hard drive.

Ready? Let's get your systems up and running on your wireless network. You'll see how step by step in this chapter, and how to avoid potential pitfalls.

Make Sure Your Hardware Is Recognized

After plugging in your equipment, check in with the Device Manager to make sure it's recognized.

1. From the Desktop, right-click My Computer and choose Properties (Figure 7.1).

2. Click the Device Manager tab (Figure 7.2).

3. From the list of devices, double-click the Network Adapters entry (or click the plus symbol next to Network Adapters in the list) (Figure 7.3).

FIGURE 7.1   Open My Computer so you can access the Device Manager.

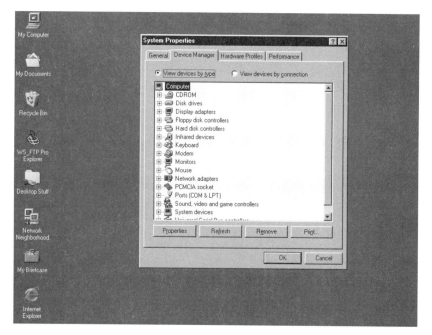

FIGURE 7.2   Choose the Device Manager tab to bring it forward.

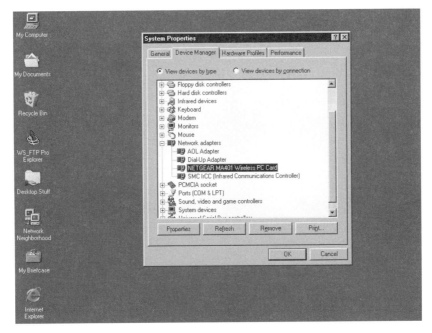

FIGURE 7.3   Double-click the Network Adapters entry.

If you do not see your wireless network adapter, you should make sure that it is connected correctly to your PC. If you still do not see the network adapter, try reinstalling the drivers. (See Chapter 6 for a list of hardware vendors and their support sites, where you can download updated drivers.)

If your equipment is recognized, it's time to see whether your computer is ready to access the Internet. If not, we'll install some software that will allow your PCs to communicate with every other computer on the Internet.

Checking for TCP/IP

First, make sure that your computer has TCP/IP (Transmission Control Protocol/Internet Protocol) installed, which will allow you to access the Internet and share files. TCP/IP, you might remember from earlier in the book, is the common language spoken by all computers on the Internet. On your home network, TCP/IP can be used both to access the Internet and to transfer files and share printers.

Okay, so let's see if you are ready to go with TCP/IP.

1. Put your operating system CD-ROM in your CD drive (you might need some files off the disk, but Windows will handle that for you).

2. From the desktop, right-click Network Neighborhood and then choose Properties (Figure 7.4).

FIGURE 7.4   Open the Network Properties dialog box by right-clicking Network Neighborhood and choosing Properties.

3. Look under the list that says "The following network components are installed:" You should see the network adapter you found in the Device Manager (from the check we made in the previous section). That means your network adapter is installed. But we need to make sure that TCP/IP is associated with your card (this is sometimes called *binding* a protocol to your network adapter).

Scroll down a bit further, and you should see "TCP/IP ->" before the name of your wireless network adapter (Figure 7.5). If you see it, you're set and can go ahead and skip to the next section. If you don't see it, see the next step.

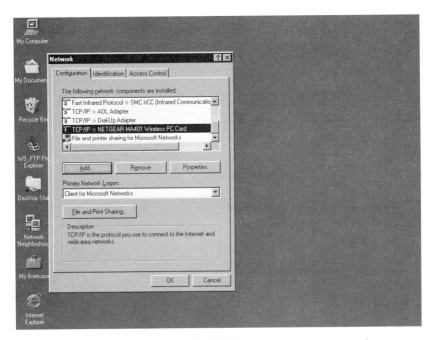

FIGURE 7.5 Check to see if TCP/IP is bound to your wireless network adapter.

4. To install TCP/IP, click the Add button.

5. Double-click Protocol (Figure 7.6).

6. Under Manufacturers, click Microsoft.

7. Under Network Protocols, choose TCP/IP (Figure 7.7). Click OK.

8. You'll be prompted to insert your Windows disk. Click OK. When prompted, reboot your computer.

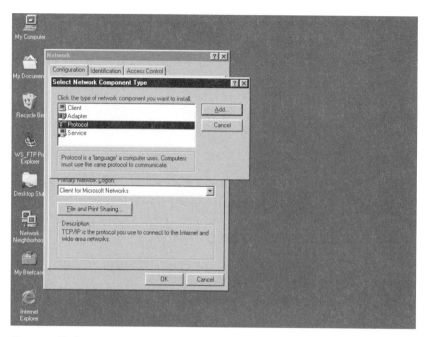

FIGURE 7.6   If TCP/IP isn't associated with your network adapter, add the protocol.

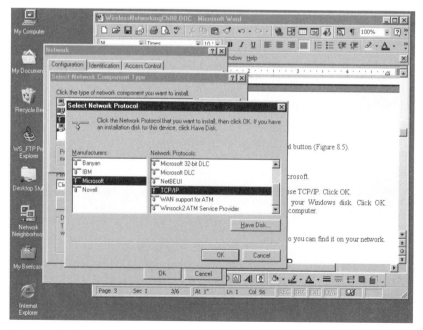

FIGURE 7.7   Choose TCP/IP.

Now it's time to name your computer so you can find it on your network.

Naming Your Computer and Workgroup

You'll need to name your computer so you can identify it (typically when you open Network Neighborhood, or if you're using Windows 2000/Me/XP, My Network Places).

1. Right-click Network Neighborhood on the desktop.

2. Choose Properties.

3. Click the Identification tab (Figure 7.8)

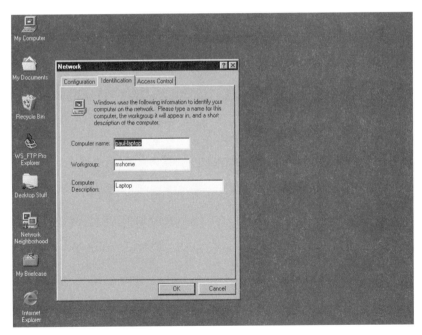

FIGURE 7.8 Click the Identification tab to bring it forward.

4. Enter a name for your computer (Figure 7.9). Choose something you'll remember.

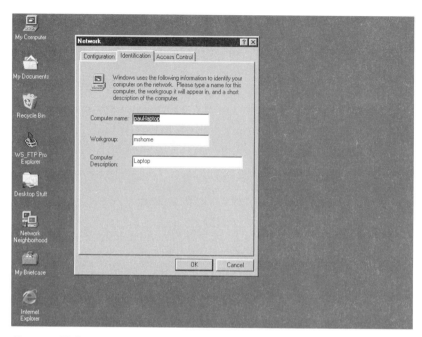

FIGURE 7.9   Name your computer.

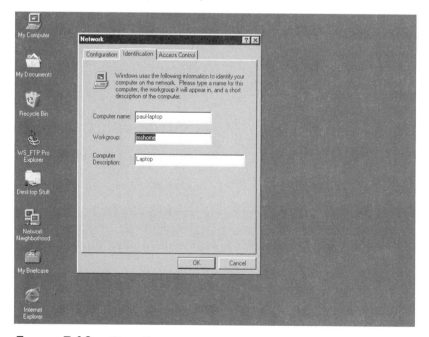

FIGURE 7.10   Give all your computers the same workgroup name. Mshome is the default name for Windows XP computers, so I use that for my Windows 98 SE computers too.

5. Enter a workgroup name (Figure 7.10). All the computers on your network should use the same workgroup name.

6. Enter a computer description.

7. Click OK.

After identifying your computers, our thoughts turn toward file and printer sharing. Windows might already be offering this service on your computer, but we need to check.

Start File and Printer Sharing

Starting File and Printer sharing is quite simple. Once you do, you should be able to view computers you share in Network Neighborhood. Keep in mind that you should always use a firewall, which hides your network from unauthorized users on the Internet. Your router, if you have one, will probably have a built-in firewall. If not, you can download a free software firewall, as we discuss in Chapter 12. With that security warning in mind, here's how to turn on File and Printer sharing.

1. Right-click Network Neighborhood.

2. Choose Properties.

3. Click the File and Print Sharing button (Figure 7.11).

4. Select "I want to be able to give others access to my files" (Figure 7.12).

5. Select "I want to be able to allow others to print to my printers."

6. Click OK.

So now you've set up your network to share files and printers, but you haven't actually shared anything yet. Let's take care of that now. Here's how to share a file:

1. Navigate to a folder you want to share, either by opening My Computer or Windows Explorer (or by selecting one from your desktop).

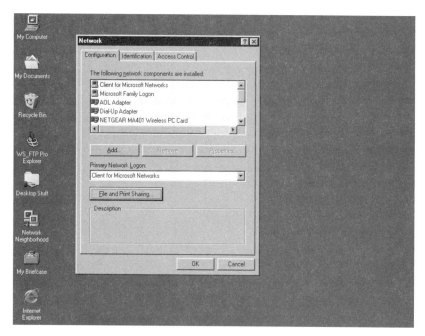

FIGURE 7.11   Click the File and Print Sharing button.

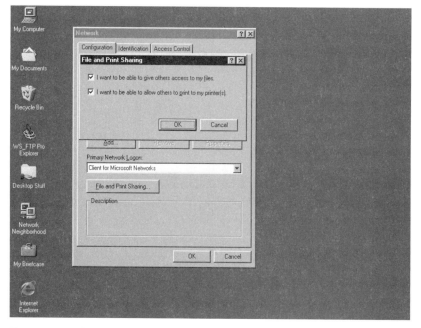

FIGURE 7.12   Start File and Print sharing.

2. Right-click the folder and choose Sharing (Figure 7.13).

FIGURE 7.13 *Choose Sharing.*

3. Select "Shared As:" (Figure 7.14).

4. Choose whether users can read or edit your files by selecting "Read-Only" or "Full."

5. If you choose "Full" or "Depends on Password," you need to choose a password.

6. Depending on whether you choose Full or Depends on Password, enter a password in the appropriate password text box.

7. Click OK.

You're all set, and your folder is now shared.

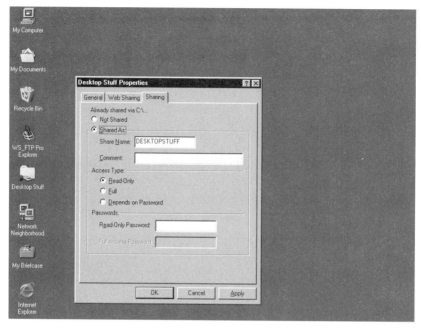

FIGURE 7.14   *Choosing Shared As will make the folder available on your network.*

What about sharing a printer? Let's share the printer connected to your PC with everyone on your wireless network. Here's how:

1. Double-click My Computer.

2. Double-click Printers (Figure 7.15).

3. Right-click the icon of the printer you want to share (Figure 7.16).

4. Choose Sharing.

5. If it's not already selected, choose the "Shared As" selection (Figure 7.17).

6. Click OK.

In the next section, we'll use Network Neighborhood on your desktop to view and use the folders and printers you've shared.

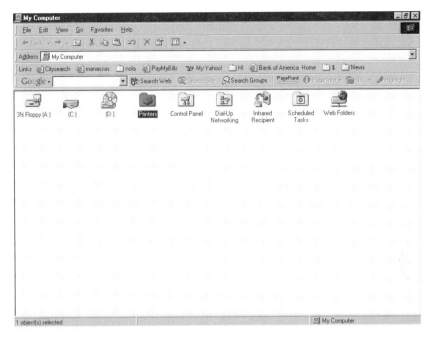

FIGURE 7.15   Double-click Printers.

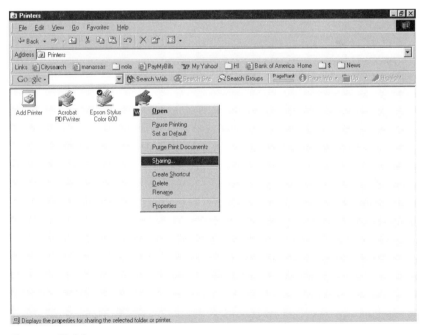

FIGURE 7.16   Share your printer.

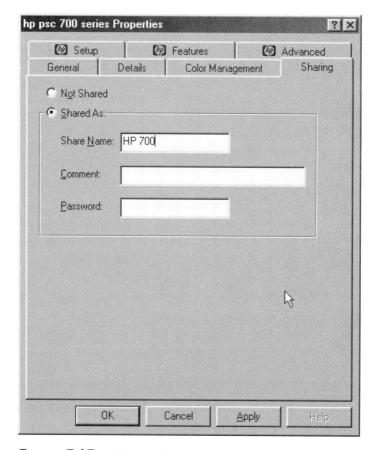

FIGURE 7.17   Choose Shared As.

Using Network Neighborhood

Finding shared resources is typically done by opening up Network Neighborhood from your desktop. If that seemed like unnecessary clutter until now, you'll start getting much more use out of it.

Just double-click the Network Neighborhood icon on your desktop to see which resources are available on your network (Figure 7.18).

You can also right-click Network Neighborhood and choose Properties to open the Network Properties dialog box. This will allow you to view the network settings for the computer you are using.

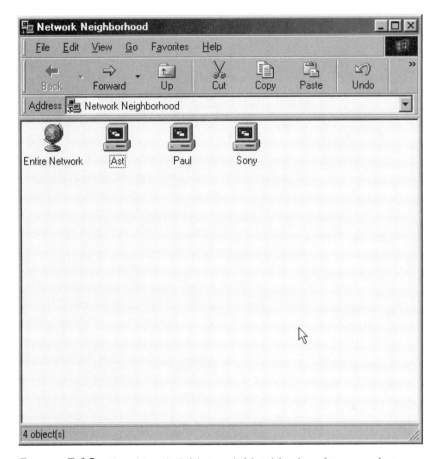

FIGURE 7.18   Double-click Network Neighborhood to see what computers are available on the network. Double-click the computer's icon to see which folders and printers are shared by that computer.

Using the Network Properties dialog box, you can turn on and off File and Print Sharing and make changes to your network adapter's settings, such as associating TCP/IP, as we did earlier in the chapter.

You can also change the name of your computer and its workgroup name by selecting the Identification tab in the Network Properties dialog box.

Share Internet Access Wirelessly

If you plan to use a computer to share Internet access, you can do so using Windows 98 Second Edition's Internet Connection Sharing software. However, your better choice is to purchase an inexpensive router (less than $100) and let the router share the Internet connection (as we detail in Chapter 10). Here's why:

If you use ICS instead of a router, you will need to leave a computer on at all times, and connected to the Internet, to share the connection.

A router will share your Internet connection without tying up a computer for hosting duties and will provide security against hackers by hiding your network computers from hackers on the Internet.

If you have Windows XP/2000 machines, check out the next chapter for the steps to connecting those computers to your network. If you have a router or want to set up Internet Connection Sharing, check out Chapter 9.

Summary

Once your hardware is connected, setting up your computers for networking is just a matter of making sure your wireless network adapter is recognized. Once it is, Windows offers all the software you'll need to connect your computers together and share data.

Networking computers wirelessly is probably less intimidating than most people think when they get started. Many computers are just waiting for the right hardware connections. Make sure TCP/IP is installed and verify that file and printer sharing is turned on. The steps in this chapter act as a checklist that should help you get every computer in your house connected.

8

Step-by-Step
with Windows XP

In this chapter...

✔ Make Sure Your Hardware Is Recognized

✔ Updating Your Drivers

✔ Change Settings Manually

✔ File and Printer Sharing

✔ Sharing Internet Access

✔ Using the Network Setup Wizard

If you use Windows XP, you're in luck. The operating system is built with networking in mind and offers features for sharing files and the Internet, as well as some built-in wireless smarts that should make setting up an 802.11 network quite simple.

You probably noticed the word "should" in the last sentence. There are a few potential hangups that we'll help you get around. Most important is to have updated drivers for your wireless network card. You can experience a range of wireless networking problems in Windows XP if you are not using the most recent driver.

We choose XP to walk you through the steps for setting up your network for two reasons. First, the operating system is Microsoft's latest and will ship with most new PCs. Second, using these directions, Windows 2000 users should be able to get up and running, since the operating systems work so similarly.

Unlike Windows 2000, XP offers the Network Setup Wizard to make the job a bit simpler. However, the Wizard, like many other things designed to make your life easier, in fact sometimes gets things wrong. It's a good idea to know how to change your network settings manually, and we'll cover that first.

Make Sure Your Hardware Is Recognized

Your first step is to make sure Windows XP recognizes your network adapter. This is straightforward. On the hardware side, make sure your wireless network adapter is plugged into the appropriate PCI port (for desktops), USB port (for desktops and laptops), or PC Card slot (for laptops). When plugged in and connected to your computer correctly, you generally see a glowing green light.

Now, you want to make sure Windows XP recognizes your network adapter.

1. First, click the Start menu, and right-click My Computer and choose Properties (Figure 8.1).

2. Click the Hardware tab. Then click the Device Manager button (Figure 8.2).

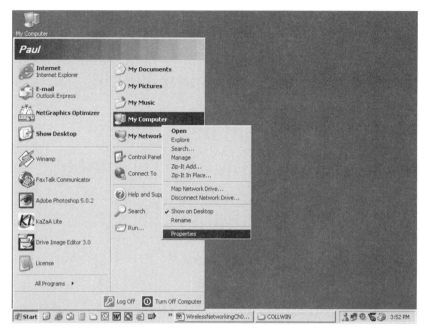

FIGURE 8.1   Right-click the My Computer icon in the Start menu.

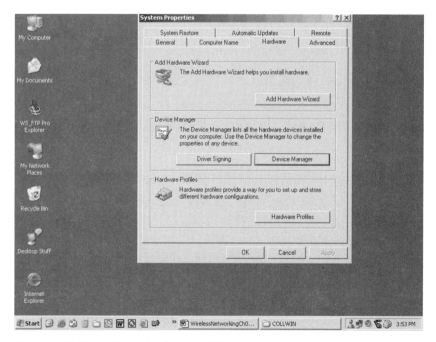

FIGURE 8.2   Click the Device Manager button.

3. In the list of devices, click the plus symbol (+) next to the "Network adapters" line (Figure 8.3).

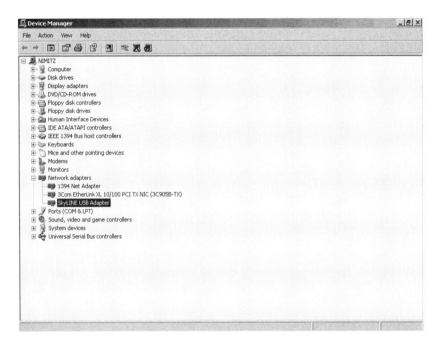

FIGURE 8.3   *Check to see if your network card is recognized.*

Your wireless network adapter should be listed here. Don't see it? Try reinstalling the driver from the CD that comes with your network adapter.

Updating Your Drivers

If your network adapter isn't working after installing your driver, you might want to look at installing updated drivers from the manufacturer's web site. Open your Web browser and head to the vendor's site, where you should be able to download an updated driver.

Below are some vendors of wireless networking equipment and their support sites, where you should be able to find the latest drivers.

> 3Com (support.3com.com)
> Compaq (www.compaq.com/support/files/networking/)
> D-Link (www.dlink.com/tech/)
> Linksys (www.linksys.com/download/)
> Proxim (www.proxim.com/support/software/)
> Netgear (Figure 8.4) (support1.netgear.com/netgear1/)
> SMC (www.smc.com)
> Intel (support.intel.com)

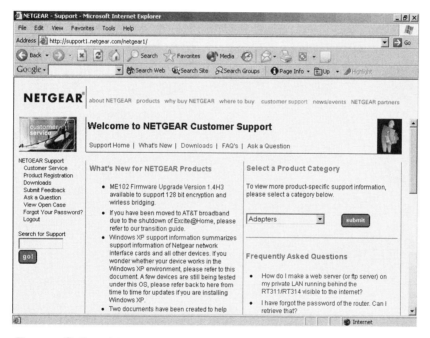

FIGURE 8.4   Netgear's support site, a good place to find the latest drivers for the vendor's wireless networking hardware.

Using an updated driver can get a network adapter up and running when you can't seem to get your computers communicating. Using the correct driver can also increase the throughput speed on your network.

Change Settings Manually

We've mentioned that Windows XP comes with a handy Home Networking Wizard, which can help you set up all the computers on your network. If you'd prefer to use the Wizard, rather than adjusting your settings manually, go ahead and jump ahead to the "Using the Network Setup Wizard" section later in this chapter. Keep in mind, though, that it's not difficult to set up your computers on your own.

First, you want to make sure your computer has a name and description.

1. From the Start menu (or your desktop if My Computer appears there), right-click My Computer and choose Properties (Figure 8.5).

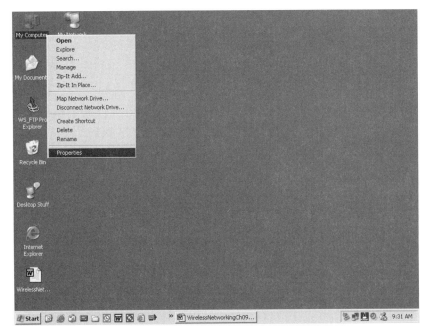

FIGURE 8.5   *Choose Properties from My Computer.*

2. Click the Computer Name tab (Figure 8.6).

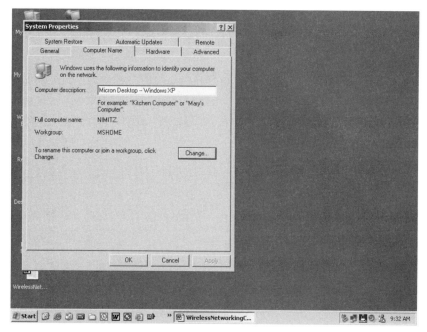

FIGURE 8.6   Click Change.

3. Enter a computer description. This can be anything you want that helps you identify the machine. You might include the machine's operating system, if you use more than one.

4. To enter a new name for the computer, click the Change button (Figure 8.6).

5. The Computer Name Changes dialog box opens. In the appropriate box, enter the computer's name. Use something straightforward or creative—your choice.

You also need to establish a *workgroup* name. When you view the computers connected to your network, they'll be grouped by workgroup. All the computers on your network should have the same workgroup name.

6. Enter a workgroup name (Figure 8.7).

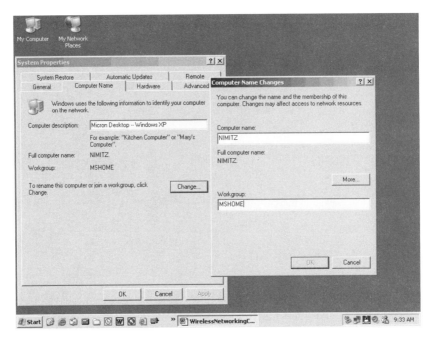

FIGURE 8.7   Enter a workgroup name, a name for your network shared by all the computers you have connected.

TIP: You can use different workgroup names, if you have a reason to, such as if you maintain several networks in your home. However, this will cause your computers to be grouped separately when you view them in My Network Places.

Once you've established your computer and workgroup name, you can move onto the real business of starting a network, sharing files and printers.

File and Printer Sharing

In Windows XP, File and Printer sharing service is enabled by default. That means once your computers have their wireless network adapters properly installed, you should be able to share files and printers.

You can check to make sure that File and Printer sharing is enabled by following these steps:

1. From the Start menu, right-click My Network Places and choose Properties.

2. Right-click the name of your wireless network adapter, and choose Properties (Figure 8.8).

3. You see File and Printer Sharing for Microsoft Networks (Figure 8.9).

4. If the box next to File and Printer Sharing (Figure 8.9) is unchecked, make sure to check it. Then click the OK button to close the open dialog box.

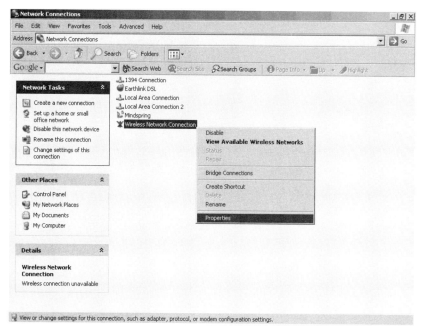

FIGURE 8.8   Right-click your wireless network adapter's icon and choose Properties.

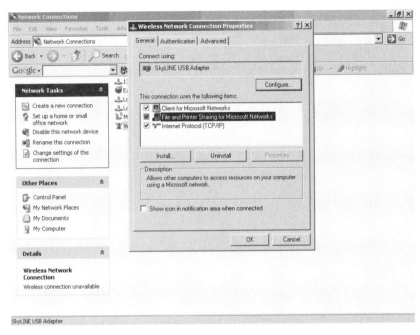

FIGURE 8.9   Make sure File and Printer Sharing for Microsoft
Networks is checked.

TIP: My Network Places works like a folder on your system
that shows you your network resources. You can view
which computers and printers are shared, and you can see
and edit the properties of your wireless network adapter.

Now that you're set with File and Printer Sharing, let's consider
sharing your Internet connection in Windows XP.

Sharing Internet Access

You can share access to your dial-up (analog 56 kilobits per second)
modem or your cable or DSL modem connection, if you have one. The
simplest way is to use the Network Setup Wizard.

If you are using a router, you can see more about sharing Internet
access through your router in Chapter 9. However, if you just want to

share a connection, and your Windows XP computer is directly connected to the Internet, just follow these steps.

1. Open My Network Places (Start, My Network Places).

2. Under Network Tasks (on the top-left side of the My Network Places menu) click "View network connections" (Figure 8.10).

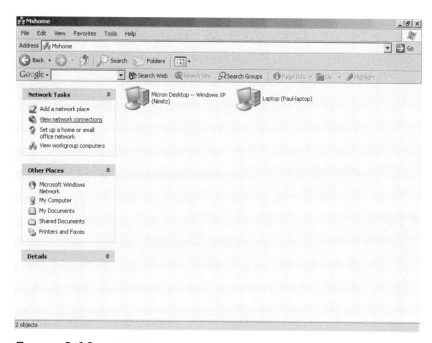

FIGURE 8.10   Click "View network connections" so that you can see your wireless network connection.

3. Right-click the wireless connection you want to share and choose Properties (Figure 8.11).

4. Click the Advanced tab (Figure 8.12).

5. Select "Allow other network users to connect through this computer's Internet connection."(Figure 8.13)

6. Select the connection you want to share under "Home networking connection."

7. Click OK.

FIGURE 8.11   Select Properties to change the settings of your wireless network connection.

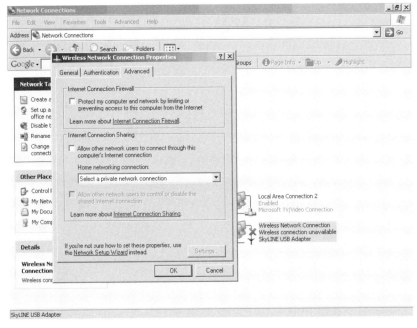

FIGURE 8.12   Click the Advanced tab to bring it forward.

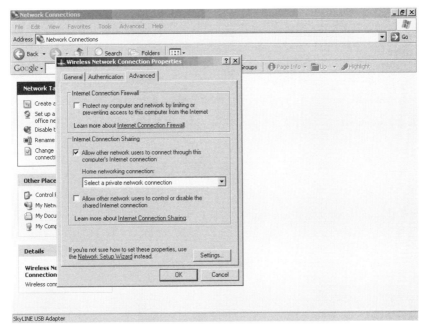

FIGURE 8.13 Set XP to share your computer's Internet connection.

Your computer is now set to act as the router for your network. Remember to leave this computer on or the other computers on your wireless LAN won't be able to access the Internet.

Using the Network Setup Wizard

If you don't want to handle your computer's network settings manually, you don't have to. The Windows XP Network Setup Wizard will tackle the job of putting your XP computer on the network in just a few minutes. No fuss, no muss. Here's how to get started.

1. Open My Network Places from the Start menu (or desktop) and, under Network Tasks, choose "Set up a home or small office network" (Figure 8.14). Alternatively, you can choose Start, All Programs, Accessories, Communications, Network Setup Wizard.

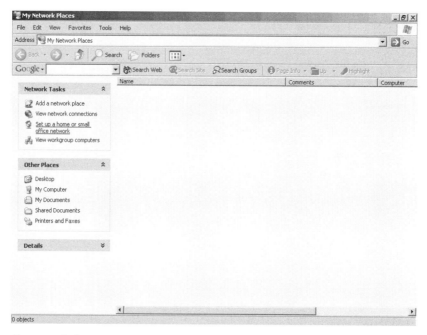

FIGURE 8.14   Start the Home Networking Wizard.

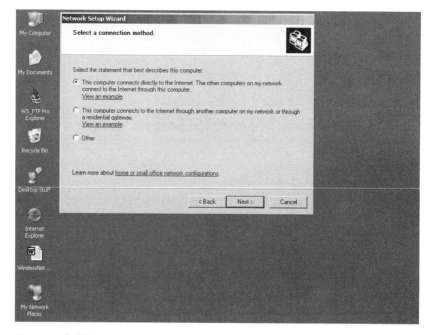

FIGURE 8.15   To share your connection, choose the first option.

2. The wizard opens. If you want to share your Internet connection, choose "This computer connects directly to the Internet" (Figure 8.15).

   If you were using a router, you'd choose the second option, "This computer connects to the Internet through another computer on my network or through a residential gateway." Click Next.

3. Choose the Internet connection you want to share (Figure 8.16). In my case, I want to share a dial-up modem connection to my ISP. Keep in mind, the Internet connection you share is probably not your wireless connection. More likely you will connect to the Internet using a wired device (a modem). You share the connection wirelessly, in the next steps. For now, click Next.

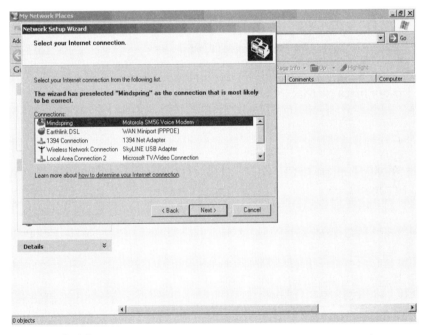

FIGURE 8.16 Choose the Internet connection you want to share.

4. The wizard can determine which network connections to share (Figure 8.17). This typically works just fine. If you want to choose the connection to share, select "Let me choose the connections to my network."

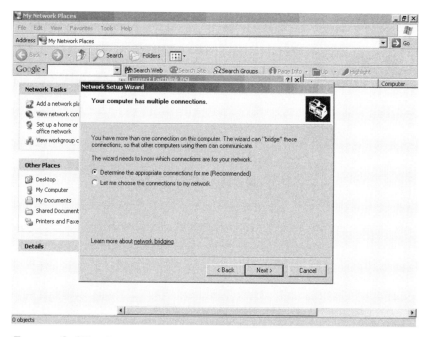

FIGURE 8.17   The wizard can determine the network connections on your machine that should be shared.

5. Choose your computer name (should be one word) and description (can be several words) (Figure 8.18). Click Next and you'll be prompted to enter a workgroup name. The default workgroup name is MSHOME (Figure 8.19). You can choose whatever name you want, just make sure it's the same on all your computers. Then click Next.

6. The wizard shows you the settings you've selected. Make sure that everything is set the way you want. If you see a problem, click the Back button and change the setting. If you're looking good, click Next and the wizard takes care of business (Figure 8.20).

When the wizard is finished, you'll be prompted to create a network setup disk to use on your other computers. You can opt out of this by clicking "Just finish the wizard" (Figure 8.21). Instead of creating a

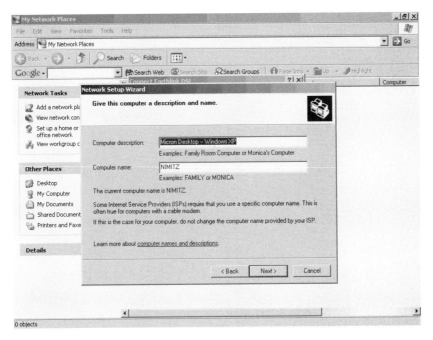

FIGURE 8.18 Choose a computer name and description.

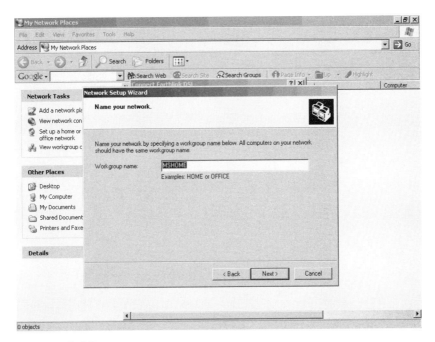

FIGURE 8.19 Choose a workgroup name.

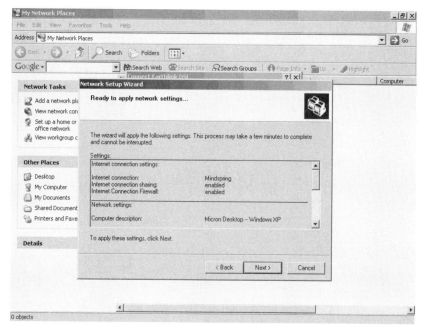

FIGURE 8.20 Verify the network settings.

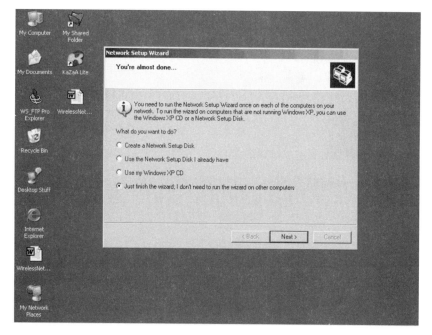

FIGURE 8.21 Finish the Network Setup Wizard.

separate disk, you can use your Windows XP CD-ROM instead. Insert it into the drive of a Windows 98 or Windows Me computer and select "Perform additional tasks" (Figure 8.22) when the splash screen appears. Then choose "Set up a home or small office network" (Figure 8.23). You can use the XP installation disk to set up all your other computers automatically (or check out the previous chapter to see how to set them up manually).

If you're having trouble after setting up your network manually or using the wizard, jump ahead to Chapter 11.

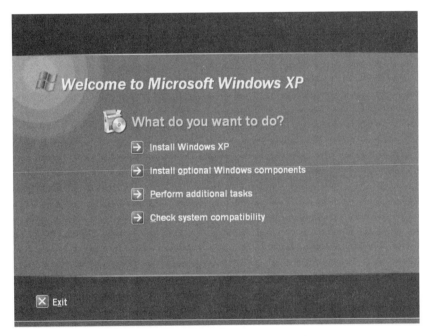

FIGURE 8.22 Choose "Perform additional tasks."

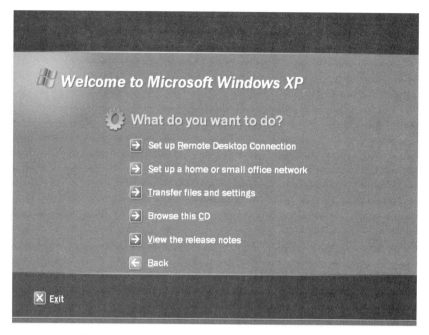

FIGURE 8.23   Click the selection "Set up a home or small office network."

Summary

Setting up Windows XP on your wireless network starts with making sure your hardware is recognized and that you have the latest drivers. You need to give your computer a name, if it doesn't already have one, and enter a workgroup name that will be shared by all the PCs on your wireless LAN. Windows XP is ready to share files and printers by default, but you should check to make sure. If you want to let XP handle these tasks for you, simply launch the Home Networking Wizard and follow along as the Wizard steps you through the process. You'll be up and running in no time.

9

Routers and Internet Connection Sharing

In this chapter...

✔ Choosing a Wireless or Wired Router

✔ Adding a Router to Your Network

✔ Setting Up Windows ICS on the Host Computer

✔ Setting Up Client Computers

Now that you know how to set up a small network, let's look a bit more at connecting your network to the Internet. Specifically, we will explore how a router, sometimes called a gateway, might make your network both safer and easier to use than if you go without one.

In addition to connecting your network to the Internet, a router provides two main benefits to your network:

1. A router helps protect your network from other people connected to the Internet through network address translation (NAT), which shows just one IP address on the Internet and allows you to use that one IP address to access the Internet from all the computers on your network.

2. A router with DHCP (Dynamic Host Configuration Protocol) handles the assigning of IP addresses to your computers so that you don't have to. These private addresses aren't visible to the Internet at large.

Most, but not all, routers provide both NAT and DHCP. Look for routers that provide these two helpful capabilities.

Could you benefit from a router? Probably. Let's consider who is a likely candidate for purchasing either a router built in to a wireless access point or a standalone wired router.

Choosing a Wireless or Wired Router

If you plan to share your Internet connection among multiple machines in your house, a wireless router will make your life easier. The cost a router adds to a wireless access point (Figure 9.1), typically less than $50, makes a combination access point and router a smart purchase.

Buying a bundled router and access point will also save you from making one more connection. Some access points include other helpful features as well, including a hub, printer server, or analog modem connection as a backup to your broadband connection. As prices fall for wireless access points with routers, manufacturers will continue adding features to remain competitive. That's good news for folks with home and small office LANs.

In the next section we consider who might be a good candidate for purchasing a separate wired router for a wireless network.

FIGURE 9.1   A wireless router from Linksys.

Already have a wired network? You can connect a wireless access point to a wired router, if you have one (Figure 9.2). Besides the differences in setup (basically connecting an Ethernet cable between the access point and wired router), there's no difference in how a standalone router or one built into an access point operates. And configuring a wired router is handled the same way as configuring a wireless one. We take a look at router setup later in this chapter.

FIGURE 9.2   You can connect a wireless access point to a wired router, such as this one from Netgear's Platinum line.

A likely scenario for needing a wired router in a home environment is the use of no-new-wires technology, such as phoneline or powerline networking hardware. Phoneline networking equipment is inexpensive and powerline equipment offers more opportunity for making network connections, so it's likely both of these technologies will remain on the networking scene for some time.

Since phoneline equipment might interest those on a budget, you could use a phoneline-to-Ethernet bridge connected to a hub, which connects your access point to all your wireless network adapters. You mix the two technologies and potentially save yourself some money, as phoneline network adapters often run less than half the cost of a wireless network adapter, and offer similar speeds.

Here are some of the possible ways to use a wired router in your wireless network:

- Connect a router with an Ethernet connection to a hub. Connect your access point to the hub.
- Connect a phoneline or powerline bridge to a hub. Connect your access point to the hub.

If you already have a wired router, there's no reason to spend the extra money on a wireless router, and your access point will be less costly. Just keep in mind that wired hardware can extend your network and allow you to make use of multiple networking technologies.

Adding a Router to Your Network

Adding a router to your network takes no time at all. Simply plug one port on your router to your broadband modem, and connect the other port to your access point (Figure 9.3). In some cases, you might not even have to configure the router's settings to work on your network. Newer routers have become so user friendly that you might be ready to go once you make all your connections.

If your router does need to be set up manually, there are several settings you should check to make sure your router is providing your computers with IP addresses, and to make sure the router is connected to the Internet.

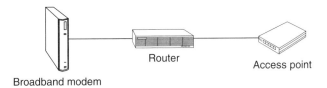

Router

Access point

Broadband modem

FIGURE 9.3   A typical modem, router, and access point setup.

It's important to know whether your ISP uses static or dynamic IP addressing. A *static* IP address is one your ISP provides you that does not change. Most ISPs use *dynamic* addresses, however, changing your address at their discretion. Dial-up modem users typically get a new IP address each time they connect to the Internet. A DSL or cable modem user might get a new IP address every few days.

If your ISP provides you with a static IP address, you simply need to enter it in your router's settings (Figure 9.4). Many routers can be accessed by opening your Web browser, and entering the IP address of the router (typically 192.168.0.1 or 192.168.1.1). The addresses 192.168.0.1 to 192.168.0.254 are subnet addresses that are only available on your network, that is, the rest of the Internet can't see them.

You can enter the IP address that your ISP provides you or select an option that will let your ISP provide the address to your router when you sign in (dynamic addressing). This should be relatively straightforward, but you'll want to check your specific router's manual to see how to access these settings.

Another setting that you might need to change, but will likely be handled for you by default, is your Domain Name Server (DNS). The purpose of DNS is to translate Internet domain names (.com or .edu addresses, for example) into unique IP addresses, which identify a computer on the Internet. A web site, for example, has a unique IP address that is associated with its domain name. DNS handles this translation. Most ISPs will pass the DNS data directly to your router, and you won't have to enter it. If you do need to enter it (your ISP will tell you), just type in the IP address of the server that your ISP provides to you. Simple as that.

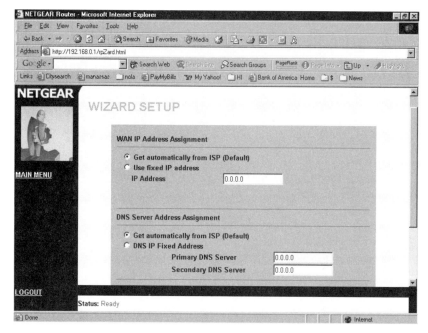

FIGURE 9.4   Your router should offer an interface that will allow you to enter a static or dynamic IP address. You might need to enter the IP address your ISP provides you.

Setting Up Windows ICS on the Host Computer

Internet connection sharing (ICS) is a Windows feature that comes with operating systems Windows 98SE and later. As we mentioned in Chapter 9, Windows XP initiates Internet connection sharing through the Home Networking Wizard. You can follow the steps in this section for Windows Me as well, which uses ICS in a similar way to 98SE. Windows Me has a Home Networking Wizard that can automate the process, though we'll walk through setting up ICS manually here.

Again, using a router is typically a simpler and more effective way to share Internet access in your home, and the setup won't vary by operating system. You won't tie up a machine that must be left on to share Internet access. And if the machine needs to be restarted (as after a system crash) or otherwise is turned off, everyone loses access to the Net. Not good.

If you are still interested in using the ICS feature of Windows 98SE or Me, here's how to do it. In most cases, the program isn't already installed, so you'll need to grab your Windows CD-ROM and place it in your CD drive.

1. Choose Start, Settings, Control Panel (Figure 9.5).

2. Double-click Add/Remove Programs (Figure 9.6).

3. Select the Windows Setup tab (Figure 9.7).

4. For Windows 98SE, double-click Internet Tools (Me users should double-click Communications instead) (Figure 9.8).

5. Select Internet Connection Sharing and click OK twice to close the open dialog boxes (Figure 9.9).

In Windows 98 SE, the Internet Connection Sharing Wizard launches. In Windows Me, the Home Networking Wizard launches. Follow the on-screen prompts to apply Internet Connection Sharing.

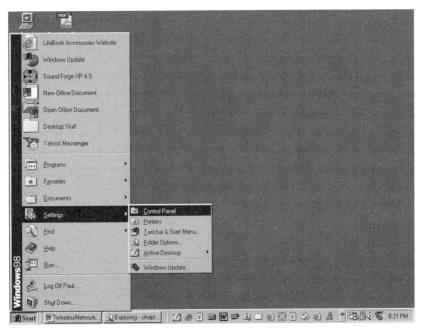

FIGURE 9.5   Open the Control Panel to get started.

FIGURE 9.6   Open the Add/Remove Programs applet.

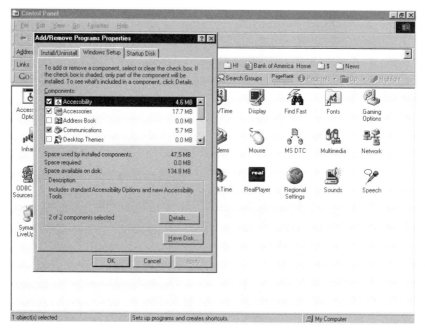

FIGURE 9.7   Click the Windows Setup tab and wait while Windows searches for installed components.

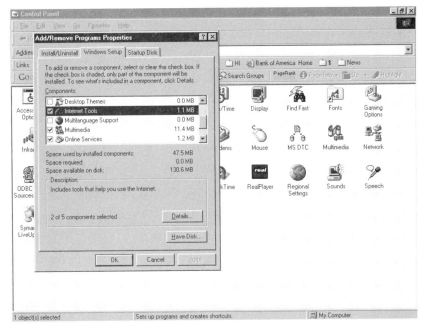

FIGURE 9.8 Open Internet Tools.

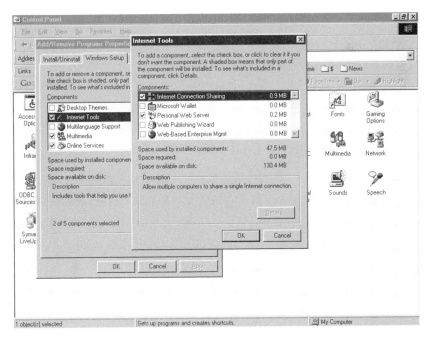

FIGURE 9.9 Enable Internet Connection Sharing.

After you finish with the appropriate wizard for starting up ICS, you can later turn ICS on and off using the Internet Options Control Panel.

1. Choose Start, Settings, Control Panel.

2. Open Internet Options (Figure 9.10).

3. Choose the Connections tab to bring it forward (Figure 9.11).

4. Under Local Area Network (LAN) settings, click Sharing (Figure 9.12).

5. Under Settings, select or deselect "Enable Internet Connection Sharing" depending on whether you want to turn ICS on or off (Figure 9.13).

FIGURE 9.10   Double-click the Internet Options icon.

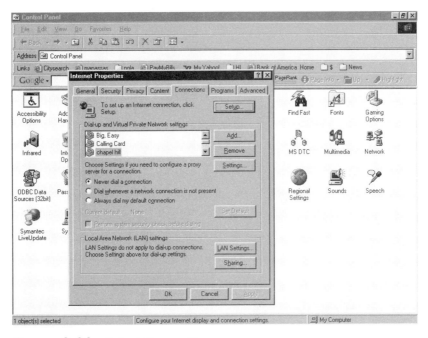

FIGURE 9.11 Select Connections.

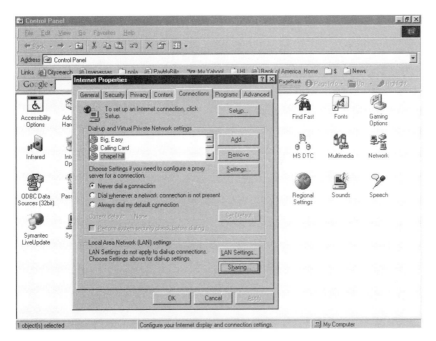

FIGURE 9.12 Click the Sharing button.

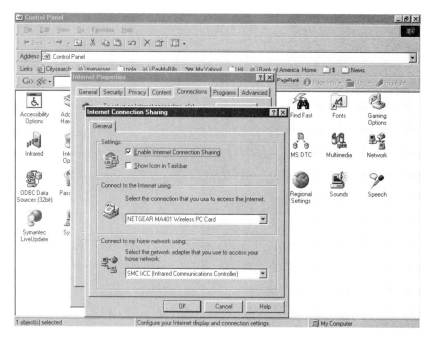

FIGURE 9.13   Enable or disable Internet Connection Sharing.

Note that if you use Internet Explorer, you can access this same menu by selecting Tools, Internet Options, and start with Step 3.

Now, one of your computers is set up to share its connection. The only thing left to do is show your other computers how to connect through the PC running ICS.

Setting Up Client Computers

During the process of using the Internet Connection Sharing Wizard in Windows 98SE (or Home Networking Wizard, in Me) you are prompted to create a Network Setup Disk. Setting up the host machines is as simple as running this setup disk on the other machines (called the clients) that will connect to the computer running ICS (called the host).

For the most part, however, you just need to make sure each machine has TCP/IP bound to the wireless network adapter in each of your machines, as we did in Chapters 8 and 9.

In the next chapter, we'll set up your browsers on your computers so that they connect to the Internet over your wireless LAN, rather than trying to dial up an Internet connection through each machine's modem. After all, you've spent considerable time and money on a wireless network. The fruits of your labor include kissing your slow 56-kbps modem goodbye (at least as long as your broadband service is up and running). Onward!

Summary

Routers are an incredibly useful tool for your LAN. Don't be intimidated if you haven't used a router before. Today's routers are so smart, you might not have to do anything more than plug the router into your network. That said, if you do need to change some settings, make sure you have the information your ISP provides you regarding connecting your computer to the Internet. With this information you can share Internet access with all the computers on your network. Because a router adds little to the price of an access point, shop for an access point with a built-in router. If you choose not to use a router, you can use a built-in feature of Windows, called Internet connection sharing (ICS), to share your connection.

chapter

10

Become
Internet Ready

In this chapter...

✔ Setting Up E-mail

✔ Browsing the Web

✔ Adding Instant Messaging

✔ Finding Your IP Address

✔ Checking Your Throughput

✔ Setting Up PPPoE

Now that you've researched and laid out your wireless network, it's time to set up the programs you'll use everyday. Sending e-mail, Web browsing, and instant messaging are just a few of the tasks that you can accomplish on your fast, new wireless network.

ere we walk through setup issues that sometimes cause headaches and look at basic maintenance that will keep your network running smoothly. We also consider setting up PPPoE (Point-to-Point Protocol over Ethernet), a step some folks will need to take for connecting their router to their ISP.

Setting Up E-mail

Setting up your e-mail accounts usually requires a few quick changes to your e-mail client. You'll need to enter your name in your client (so that recipients will recognize you), set up your user name and password, and enter the server addresses where you pick up your mail. We'll look at these steps in the popular Microsoft Outlook Express program. These basic steps are the basis for setting up any e-mail program to pick up mail over your network.

1. First, set up a new account. Choose Tools, Accounts (Figure 10.1).

2. Choose Add, Mail (Figure 10.2).

3. The Internet Setup Wizard launches (Figure 10.3). Enter the name people will see when they receive your mail. Click Next.

4. Enter your e-mail address. Click Next (Figure 10.4).

5. Now enter the server addresses your ISP provides to you (Figure 10.5). The mail server receives and stores your incoming mail. The SMTP (Simple Mail Transfer Protocol) server is your outgoing server, the one that will handle all the mail you send. If you are going from a dial-up account to a broadband Internet connection, check out the tip on p. 144, as you might need to use your old SMTP server to pick up mail from your dial-up account. Your

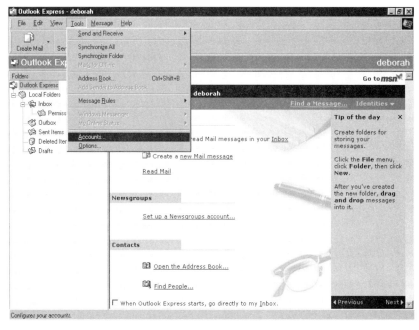

FIGURE 10.1 Set up a new e-mail account.

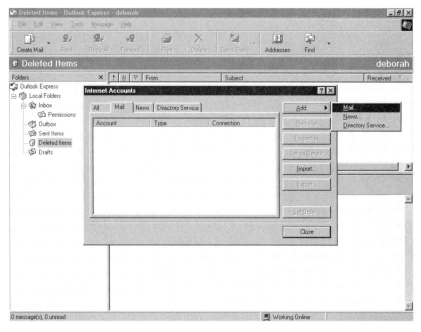

FIGURE 10.2 Choose Mail to start the Internet Connection Wizard.

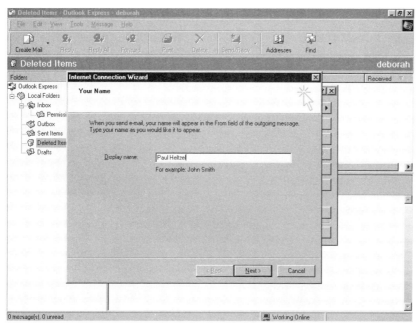

FIGURE 10.3 Enter your Display name.

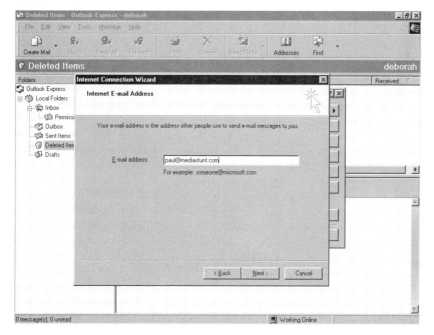

FIGURE 10.4 Type in your e-mail address.

broadband service provider will include an e-mail account with your service, but you might not want to do away with a dial-up account you've had for some time, or one that you use when traveling.

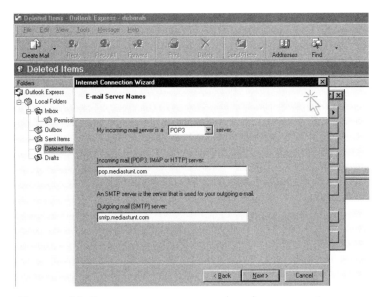

FIGURE 10.5 Add your incoming (mail) server and outgoing (SMTP) server.

6. In the Account name text box, enter the username your ISP provides you (usually the part of your e-mail address before "@", although some providers use your entire e-mail address to log in). You also need to enter your password. Select the "Remember password" option so you don't have to enter it each time (Figure 10.6).

7. That's it! Click the Finish button to complete the setup (Figure 10.7). If your Internet connection is active, you can press the F5 button (the shortcut for Send/Receive mail).

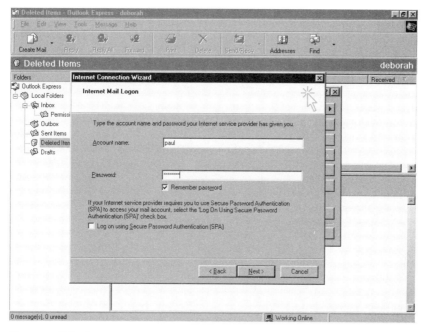

FIGURE 10.6 Enter your account name and password.

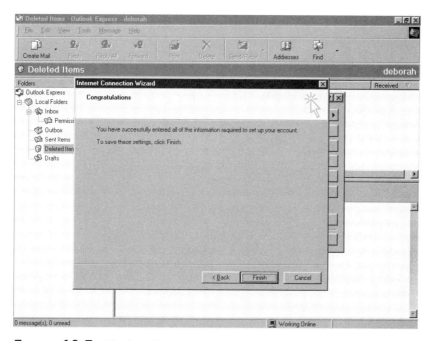

FIGURE 10.7 You're all set.

TIP: If you've been using a dial-up account but will now share a broadband modem over your wireless network, you'll probably need to mix the mail server from your old account, and the SMTP server from your broadband account. The catch? As a security measure, some ISPs won't allow you to access their SMTP (outgoing) mail server unless you are also connecting through that provider's Internet connection. For instance, to use the e-mail account from your dial-up connection, you will need to use the mail server from your dial-up account, but use the SMTP server of your broadband connection. Some providers allow this, some don't. You won't know until you try it. If your provider won't allow you to use the SMTP server from your dial-up connection over your broadband account, you can probably still pick up your mail and send it using your dial-up ISP's Web site (most ISPs provide Web-based mail for folks to pick up mail when they're on the go).

Now that you're ready to send and receive mail, let's consider how to start browsing the Web over your wireless network.

Browsing the Web

As in the last section, here we look at moving from a dial-up Internet connection, where your computer connects directly to the Internet, to a shared connection over your wireless network. Each computer accesses the Internet through a router or a computer directly connected to the Internet, sharing its connection through Windows Internet connection sharing (ICS).

You can use Internet Explorer to let your computer know that it should not attempt to dial a connection when you use your Web browser.

1. Choose Tools, Internet Options (Figure 10.8).

2. Click the Connections tab (Figure 10.9).

3. Choose "Never dial a connection." You can also choose "Dial whenever a network connection is not present" if you also have a modem connected to your PC. This might be helpful to folks who want to use their modem as a backup in case their broadband connection temporarily fails.

Next we'll look at the software all the kids are raving about: Instant Messaging, or IM.

Adding Instant Messaging

Instant messaging (IM) is a simple chat tool that can be really handy to use over your wireless home or office network. For quick communication at home or in an office, you don't necessarily need to send an e-mail. And your wireless network creates a perfect way for you to exchange quick messages. Example:

"Bob, I'm @ the pool, using my cool wireless connection. Bring me a soda."

or:

"Lucy, I'm on the phone. Pick up line 2."

IM clients require an Internet connection, but now that you're sharing a connection over your wireless network, every PC can have its own IM software, and you can use the software to chat with each connected person in your home or office.

Windows XP ships with an IM client, Windows Messenger (Figure 10.10). There are many other popular IM clients, including the programs in Table 10.1.

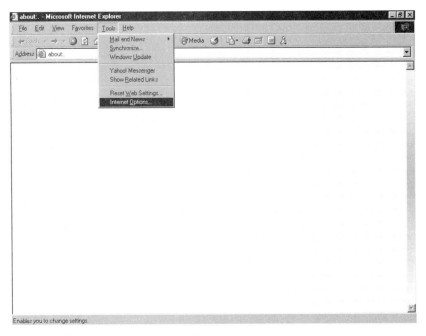

FIGURE 10.8 Open Internet Options.

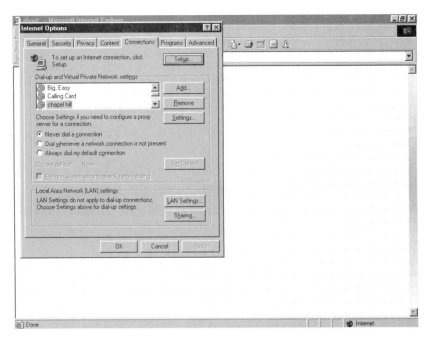

FIGURE 10.9 Tell your computer that you're on a LAN, so it won't try and connect using your PC's modem.

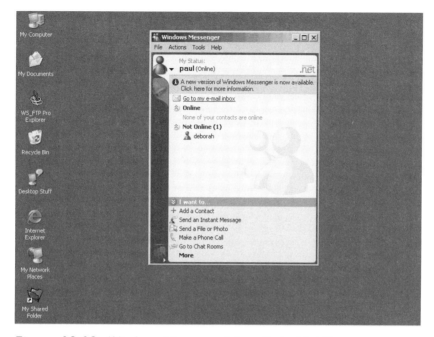

Figure 10.10 Windows Messenger is bundled with XP.

TABLE 10.1 Instant Messaging Clients and Where to Find Them

IM client	Download site
Windows Messenger	messenger.msn.com
AOL Instant Messenger	www.aim.com
ICQ	web.icq.com
Yahoo Messenger	messenger.yahoo.com

You set up each of these IM programs in basically the same way. (Some steps might vary in order a bit by software maker, but here's the basic idea.)

1. Download the IM client (Figure 10.11).

2. Install the program.

3. Register for your account. Log in and enter the user name and password you chose at registration (Figure 10.12).

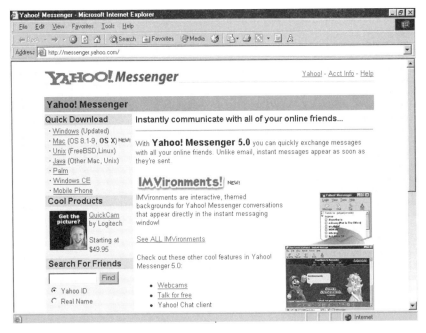

FIGURE 10.11   Downloading Yahoo Messenger.

FIGURE 10.12   Log in to Yahoo Messenger.

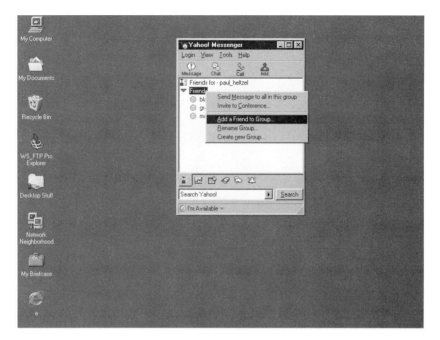

FIGURE 10.13 Your buddy list.

4. Enter the user names of people you know to set up your "buddy list" (Figure 10.13).

You can now send messages to anyone with an IM account. Note that most IM programs can only communicate between users of the same software (Yahoo Messenger users, for example, can't send instant messages to AOL Instant Messenger users).

Finding Your IP Address

You might find that you'll need to know your IP address on a machine in your wireless network, for instance, when you play a network game or, later in this chapter, when we use a program called Qcheck to check how fast your wireless network transfers data.

To find your IP address in Windows 98/98SE/Me:

1. Click Start, Run.

2. In the Open box, type `winipcfg` (Figure 10.14).

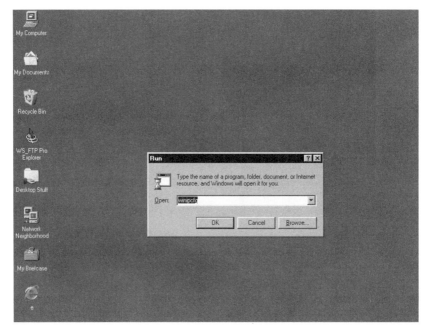

FIGURE 10.14 Open winipcfg.

3. You should see the IP address listed. If you do not see your IP address, click the pull-down menu (Figure 10.15) and choose your wireless network adapter.

   In Windows 2000/XP:

1. Choose Start, All Programs, Accessories, Command Prompt (Figure 10.16).

2. Type `ipconfig` (Figure 10.17).

You'll see the IP address of the network adapter you are using.

Now let's use that IP address to see how fast your wireless network really is.

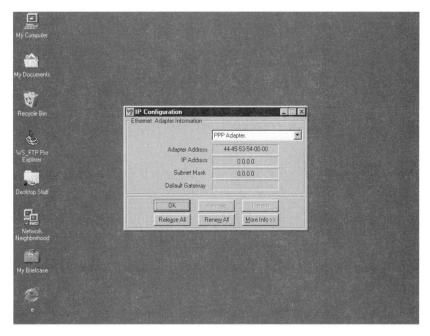

FIGURE 10.15 Check your IP address. Here the menu is showing the PPP (dial-up modem connection).

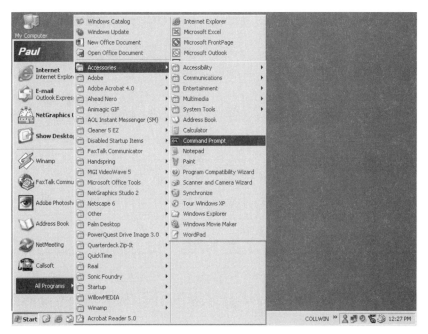

FIGURE 10.16 Open the Command prompt.

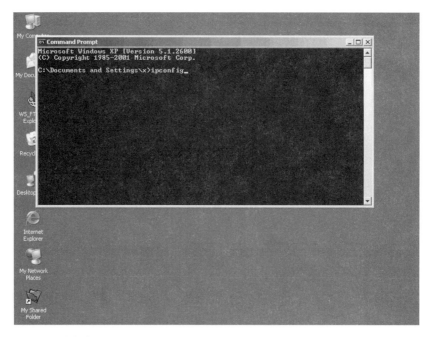

FIGURE 10.17 Open ipconfig.

Checking Your Throughput

The rate that data passes over any network, wireless or wired, can vary greatly depending on several factors. There's overhead: the chatter between computers to prepare for and confirm the data transfer. Then you have the distance over which the data travels. As you know, the farther you are from your access point, or another wireless network adapter, the slower your data transfer will be.

You can check the rate of data transfer by downloading a handy, free program called Qcheck and then installing it on (at least) two machines that you want to test. You install the program and launch it on one machine, then run the program on another machine. Using a simple, friendly interface, you click a button to send sample data packets over your network, which Qcheck times to see how fast your network is communicating.

You might be surprised that your wireless network is sending data at less than half, sometimes much less than half, its maximum speed. In my home, my 11 mbps 802.11b network tends to send data no faster than 5 mbps. My 802.11a network (with a maximum possible data rate of 54 mbps) sends data at about 22 mbps. These are the speeds I see in ad-hoc mode. In infrastructure mode (using an access point), they are about half these speeds. In most cases, it makes little difference, since your cable or DSL modem likely transfers data at a maximum of 1.5 mbps. The slowdown hurts the most when transferring files from one networked computer to another.

To use Qcheck, follow these simple steps:

1. Download the program from www.netiq.com.

2. Install the program on one computer on your network (Figure 10.18).

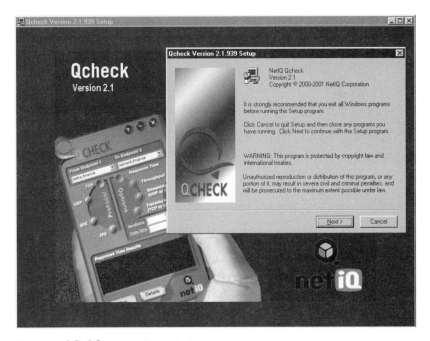

FIGURE 10.18 Installing Qcheck.

3. Install the program on at least one other computer on your network.

4. Launch the program on both your computers (it needs to be running on at least two computers).

5. Click the Throughput button to start the speed test (Figure 10.19).

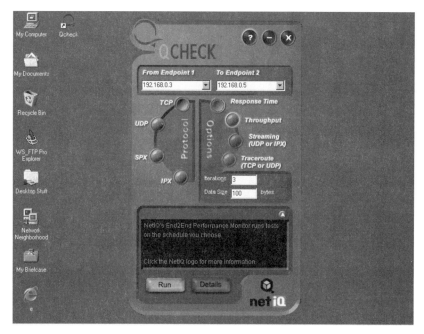

FIGURE 10.19   Click the Throughput button.

Qcheck sends data packets between the two computers and records the time it takes for the packets to transfer. Click the Throughput button several times to run at least three tests. Running multiple tests and averaging the results should provide you with a reliable idea of the speed at which the data is transferred over your wireless network.

Setting up PPPoE

Some broadband Internet providers use a technology called Point-to-Point Protocol over Ethernet (PPPoE) which provides secure access to your broadband connection. The added security typically means a little more work for you.

> **TIP:** Point-to-Point Protocol (PPP) is a common protocol used for dial-up connections to the Internet. If you use a dial-up modem to access the Internet, you are probably doing so over PPP.

A handy feature most home network routers use is the ability to set up PPPoE once, so that you won't have to sign in each time you want to use the Internet. All you need to do is enter a username and pass-

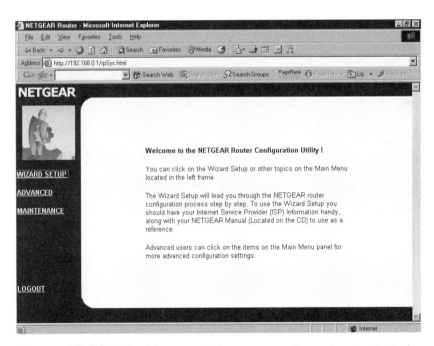

FIGURE 10.20 This Netgear wireless access point and router includes a wizard menu that will walk you through basic setup, accessed through a Web browser.

word, which will authenticate you as a user and make the connection. Routers vary slightly by manufacturer, but here's the general idea, using a Netgear MR314.

This router offers a wizard menu (Figure 10.20) that takes you through the basic setup of the hardware. One menu screen offers a form in which you can provide the PPPoE information that your ISP gives you (Figure 10.21).

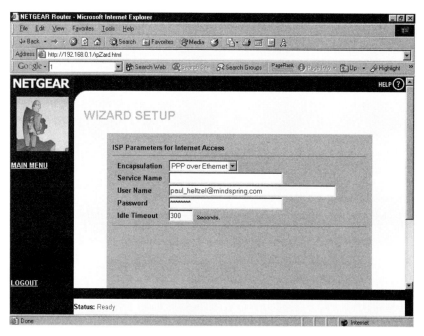

FIGURE 10.21   To enable PPPoE, you enter your username and password, which your ISP provides you.

Not all routers support PPPoE. So if your broadband ISP uses PPPoE, make sure to buy a wireless access point/router or standalone router that offers support for PPPoE—many do. Happy surfing!

Summary

What's the use of a wireless network without handy Internet software? If you're used to dialing up your ISP over an analog 56-kbps modem, you're probably going to need to make some changes to the way your e-mail client and Web browser connect over your network. If you haven't used instant messaging, you might find that your shared Internet connection offers a good reason to try chat software that can be used for fun and profit. The rate at which your network transfers data can vary greatly, so consider using Qcheck, an effective way to gauge network speed. Finally, if your ISP uses a feature called PPPoE, you can set up your router to handle this task automatically in order to share your connection without logging in from each computer on the wireless network.

11

Troubleshoot Wireless Connections

In this chapter...

No matter how well you plan out your wireless network, you're bound to come across problems. These problems typically, frustratingly, seem to come out of nowhere, and they often at first blush have one thing in common: Your network isn't working and you haven't changed a thing.

Wireless network problems often resemble one of the following scenarios:

- Yesterday everything was working perfectly, and now nothing works.
- One machine is configured just like all the others, but it still won't connect.
- One computer has Internet access, but it can't be viewed by the other PCs on your network.

Most often, the problem is simple and you'll be up and running in no time. If you make sure you have up-to-date drivers (no small matter), make sure all machines have the same configuration settings (using the software that comes with your network adapter and in Windows) and make sure your connections are secure, you should find the problem, as long as you're patient and cover all your bases.

The first step is to forget what you know. Challenge the most basic assumptions, like whether your equipment is getting power or checking settings that you *just know* haven't changed. Then check out the tips in this chapter.

Are Your Drivers Current?

An outdated driver can sink your network before you even get started. Sadly, with both new wireless networking hardware and new operating systems always being introduced, there are too often incompatibilities.

Probably the biggest problem users will see with wireless networking and driver problems occur when a new operating system does not support older equipment, or supports only some features of this equipment.

When Windows XP was first introduced, and even today, some wireless network adapter manufacturers have not released drivers that fully support their hardware. Many vendors simply put out their Windows 2000 drivers because the two operating systems are similar (Windows XP is based on Windows 2000 technology and despite being a "consumer" operating system, much more closely resembles this OS than Windows 95/98/Me.)

Some Windows XP users with 802.11b still can't get anything except the Windows 2000 drivers for their hardware. Windows XP does not recognize some of these drivers and does not make use of all the features you should get out of your card.

The bottom line? Make sure the hardware vendor you plan to purchase wireless networking hardware from offers a driver for the OS you use, whether that OS is older or the latest available. Check the vendor's Web site before you buy.

Finding A Bad Network Adapter

If your drivers are current, and you can't seem to get a connection using your wireless network adapter, you might have a bad radio or other hardware malfunction.

Try using the malfunctioning wireless network adapter on a PC that has a working network connection and wireless network adapter. If you still can't get a connection on a USB model, try replacing the USB cable.

If you still can't get the adapter to function, you might need to send it back to the manufacturer. Try swapping in a working unit, however, and you'll know for sure.

The Trouble with a Mixed Windows 98 and 2000 Network

Many people find when they mix Windows 95/98 and Windows 2000 operating systems, problems ensue. The biggest difference in these operating systems is that Windows 2000 offers a level of security that the others do not. This can often lead to trouble connecting Windows 95 and 98 computers to Windows 2000 PCs.

The problem might look like this example: The Windows 2000 computer can see the shared files and printers of a Windows 98SE machine, but not the other way around. The reason? Windows 2000 requires that you set up a user name and password for each person that will connect and share its resources. This is often frustrating because the earlier (95/98) machines connect so easily together, and for that matter so do Windows XP computers.

Here's how to make the Windows 2000 machine ready to share well with others:

1. Choose Start, Settings, Control Panel, Open Users and Passwords. Click the Add button (Figure 11.1).

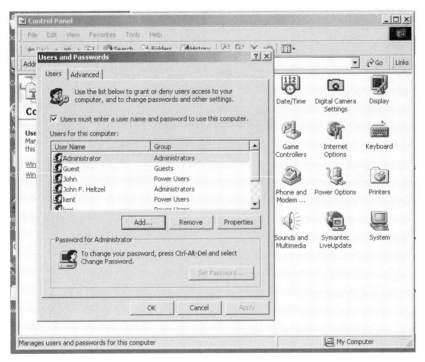

FIGURE 11.1   Open Users and Passwords to start adding a new user.

2. Enter a name and click Next. Enter a password in the next screen and click Next again.

3. Choose whether the user will be able to install programs on this computer. You can choose Restricted User to keep users from installing programs on this computer. In most cases, you can just leave Standard User selected. Click Finish (Figure 11.2).

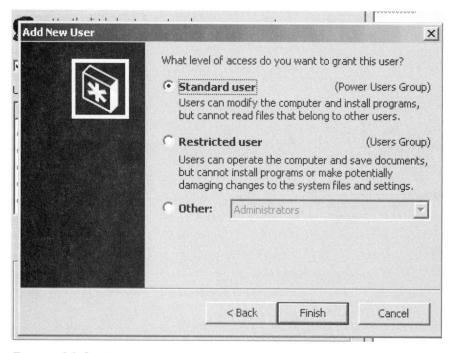

FIGURE 11.2 Choose whether a user on your network can install programs on this computer.

4. Windows 95/98/Me users will need to log in by choosing Start, Login, and entering the name and password you have given them at the Windows 2000/XP machine. Access won't be allowed unless they use the name and password you set in Step 2.

Your Windows computers will be able to view each other on the network and share files. Now we'll consider some wireless-specific network connection problems.

Configuration Utility Errors

When your wireless network won't work, first check the configuration utility settings for each network adapter. Get one slightly off from the rest, and it won't establish a link to the wireless network.

As mentioned earlier in the book, three settings provide most of the problems in an 802.11b network. Check these first when you're having trouble:

1. Make sure your network is set to ad-hoc, if you don't have an access point, or to infrastructure, if you do use an access point.

2. If you are using encryption, make sure it's on and at the same level on all the machines and on the access point (if you have one) on your network (Figure 11.3). If one wireless network adapter is set to 64-bit encryption, and another is set to 128-bit encryption, they will not communicate.

3. Check to see if your Service Set Identifier (SSID) is the same on all machines (Figure 11.4). Like encryption, if you have the wrong setting on one machine, it won't connect to the wireless network.

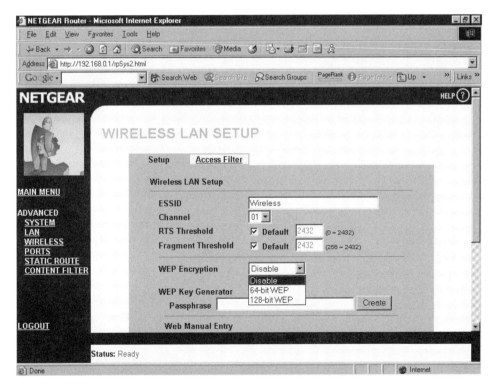

FIGURE 11.3 Make sure encryption is disabled, or set to the same level at all your computers and at your access point.

FIGURE 11.4   Make sure your SSID is the same on all machines.

You can use the SSID **ANY** on many network adapters, which will allow the use of any SSID to connect.

Once your configuration software for your network adapters is set correctly you should be all set. There is, however, another potential software setting you can check if your network won't communicate. We'll look at software firewall settings next.

Firewall Hangup

Firewalls can sometimes do their job a little too well. As we mentioned earlier, a firewall helps keep people without authorization off your network. At the same time, your firewall, especially the built-in one in Windows XP, as well as other software firewalls that run on your computer, can keep *you* off your network if you don't set it up correctly.

First, consider whether you need a firewall installed on each machine on your network. Most people choose either a hardware router with a built-in firewall, or add a software firewall to their computers. If you have a hardware router with a firewall, you don't necessarily need a software firewall on each PC. The firewall in the router should provide enough protection to keep hackers or other interlopers from gaining access to your network over the Internet.

If you use Windows XP, you can turn off the firewall. Here's how:

1. Open My Network Places (Figure 11.5).

2. Right-click your wireless network connection and choose Properties (Figure 11.6).

3. Click the Advanced tab (Figure 11.7).

4. Deselect "Protect my computer and network by limiting or preventing access to this computer from the Internet"

5. Click OK to close the open dialog boxes.

If you use a personal firewall, such as the free (for personal use) ZoneAlarm from Zone Labs (www.zonelabs.com), you can set the software to recognize computers on your wireless network. Download the program and follow the installation instructions. Launch the program and follow these steps to allow it to work on your network:

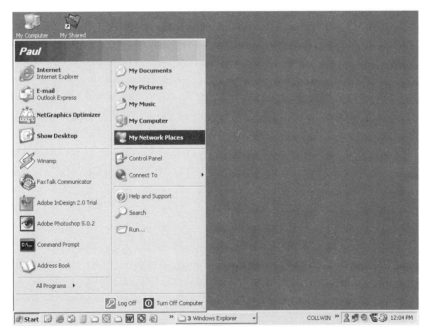

FIGURE 11.5 Open My Network Places.

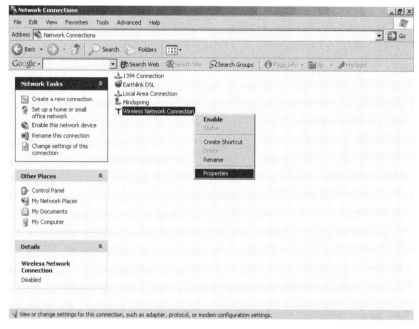

FIGURE 11.6 Open your wireless network adapter's properties.

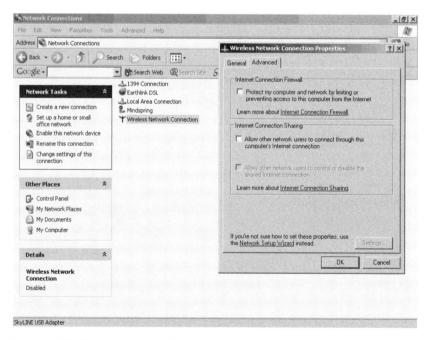

FIGURE 11.7 If you use a firewall built into your router or router/wireless access point, you can turn off the firewall in Windows XP.

1. Click the Security button (Figure 11.8).

2. Click Advanced (Figure 11.9).

3. Select your network adapter (Figure 11.10)

4. Click Add, then click the Apply button.

This firewall connection hangup is *very* common, especially when you add a new computer to your network. Check to make sure the firewall settings are set correctly.

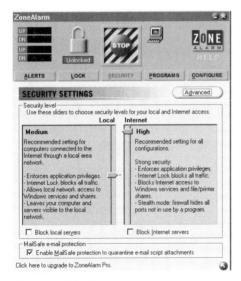

FIGURE 11.8 Click the Security button.

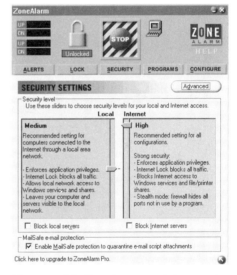

FIGURE 11.9 Click the Advanced button.

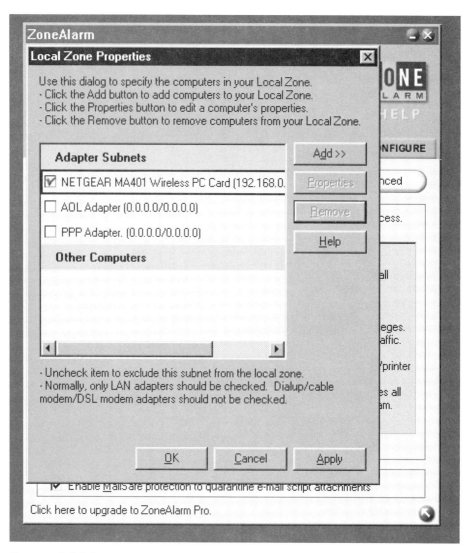

FIGURE 11.10 Choose your wireless network adapter.

Printer Problems

Problems using a printer often happen in mixed operating system environments, when for some reason a driver is corrupted by other software installed on a computer.

The best bet, as in many other situations, is to make sure you have the latest driver for your operating system. Each computer needs the correct driver installed to access a shared printer.

If a printer does not appear in Network Neighborhood (in Windows 95/98) or My Network Places (Windows Me/2000/XP), make sure the user is logged on at the PC that is attempting to share the printer.

Other Common Problems

Still having problems? Here are a few quick tips to consider:

- Try testing all your wireless equipment in close proximity to rule out problems related to the range of your wireless networking equipment.
- Turn off wireless encryption (WEP) while setting up your equipment. Turn encryption back on once you have made a connection.
- Can you access the Internet? If you can, but can't access computers on your network for file and printer sharing, you likely have a network settings problem in Windows. Check the settings of a computer that's connecting and make sure the settings are the same for the computer that can't connect. No access to the Internet? Check your broadband modem: Do you have a link light that shows whether your ISP is up and running? (Most broadband modems have three lights that should be lit continuously: One light for the connection to the ISP's network, another that shows a connection to your network, and one that shows the modem is getting power.)

Finally, when all else fails, reboot—everything. First the modem, then your router (if you have one), the access point, and finally, your computer. Make sure each device is up and running before you reboot (a.k.a. power cycle) the next.

Summary

Everybody runs into a few problems connecting their network, especially the first time around. The good news: Once you get the kinks worked out, you'll likely not need to worry much about your wireless network. The bad? Occasionally, you run into a problem, such as a malfunctioning radio in a wireless network adapter, that is hard to diagnose and will stop your network from running.

chapter

12

Security

In this chapter...

✔ Firewalls

✔ Choosing a Hardware or Software Firewall

✔ Wireless Encryption

✔ Router Filters

✔ How to Encrypt Your E-mail and Other Data

✔ Viruses and Your Network

Security might be the last thing on your mind when setting up a wireless network. And while you might not see why anyone would want to go mucking around in your personal files, you don't want to leave your computers completely unsecured. As a general rule, you should employ a level of security that corresponds to the sensitivity of the data on your network.

Some reasons to secure your computer:

- Your personal financial data and medical history might be on your hard drive.
- Fast Internet connections aren't cheap, and you don't want to share your bandwidth with anybody who can drive by your home or office.
- You don't want someone churning out spam from your PC or from your e-mail address.

Wireless security is inherently easier to crack than wired networks because there's no need for a physical connection to your network. Data transfer occurs over the airwaves, and that makes accessing it easier.

You can take a few precautions to increase the level of security on your network. These tips are quick and painless, and they'll make your data safer.

Firewalls

We've mentioned firewalls throughout the book. In most wireless LANs, the firewall is built into the access point, which also contains a router. In most small-office and home situations, a firewall can help keep unwanted users out of your network, and you shouldn't have to give it much thought. Here we'll explain the two basic ways that firewalls work so that you can choose which one you would like to use (you really should use either a hardware or software firewall).

When we say a hardware firewall, we're actually talking about software running on a hardware router. You can, in fact, purchase an expensive, stand-alone hardware firewall, but this level of security is typically necessary only in a large business network.

Most home and office wireless networks will be well served by the firewall built into your wireless access point/router (Figure 12.1).

FIGURE 12.1 This wireless access point from SMC has a built-in firewall.

Choosing a Hardware or Software Firewall

Look for a hardware firewall that provides Network Address Translation (NAT). NAT hides your network from outsiders by handling requests on your network for data from the Internet then passing it on to the correct computer within your network (but not letting any requests come from outside your network).

A software firewall, such as those made by McAfee, Symantec, and Zone Labs, can also keep data from going out of your network onto the Internet (Figure 12.2). Software routers are sometimes called personal firewalls because they are installed at each machine (you don't need a network to use a personal firewall).

Why would you want to scan requests for data from *within* your network? Some viruses that work their way onto your computer can do their worst damage once they start sending information from your computer, which then allows them to continue to spread as recipients open e-mail attachments they see as a trusted source—namely, you.

Software firewalls can offer protection against e-mail viruses, which send data from your computer over the Internet (Figure 12.3).

Software firewalls can also alert you to "spyware" activity on your computer. *Spyware* is software downloaded onto your computer, sometimes without your knowledge, which connects to its maker and

Security

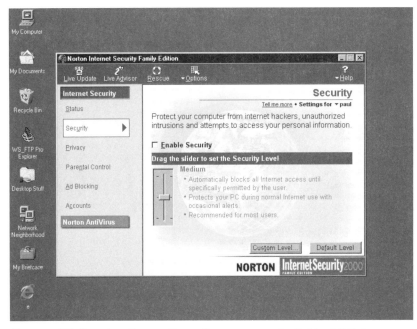

FIGURE 12.2 A software firewall protects an individual PC from unauthorized access over the Internet.

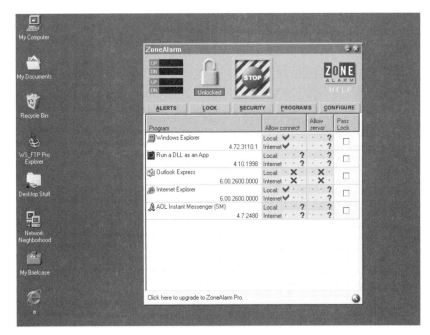

FIGURE 12.3 Zone Lab's Zone Alarm can keep software—or a virus—on your computer from contacting the Internet without your permission.

reports back information from your PC. Spyware is often bundled with software you download, and installs without your knowledge. Typically, spyware can't identify you uniquely, but the software will attempt to contact its maker to collect aggregated user information. With a software firewall, you get to decide what programs can send data from your computer over the Internet, even if they are transparently working in the background.

The Internet security suites offered by McAfee and Symantec, combine antivirus software with personal firewalls and also offer *intrusion detection (ID)*. Intrusion detection can alert you to the presence of a hacker who might be deleting files, for example, or accessing directories in the background without your knowledge.

ID acts as a second line of defense, for instance, if your antivirus software did not catch a Trojan horse. A Trojan horse is a program that appears to be safe, or useful, but instead installs malicious code on your system. A Trojan horse could allow a hacker to control your computer surreptitiously over the Internet. If you should suffer a Trojan horse, your ID software can alert you that actions are being taken by your computer that you did not authorize.

Wireless Encryption

Wireless encryption can help make your network safer by scrambling data sent by your wireless network adapter so other wireless network users won't accidentally or purposefully connect to your network.

WEP (wired equivalency privacy) is used on wireless networks to secure data. You choose a string of text, called a *key*, that is combined with the data sent over your network to make it indecipherable to anyone who does not have the key.

Enabling encryption is quite simple. You need to set up encryption by changing a setting at both the network adapter and the access point, if you use one. Then you enter the same passphrase, which generates a key (sometimes called a *pass key*) at each computer and at the access point, using the software utility that comes with your wireless networking hardware (Figure 12.4).

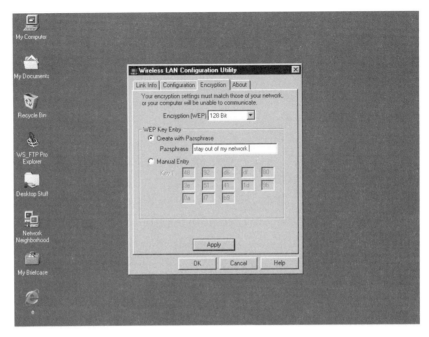

FIGURE 12.4   Enter your passphrase using the software that comes with your access point and on your network adapters.

WEP has come under considerable scrutiny because a smart hacker can likely beat the encryption method if he is determined to do so. This fact makes wireless encryption inherently more insecure than wired networks. However, for keeping your neighbor or the office next door off your network, WEP should be fine. In the next section we look at other methods you can use to make your network more secure.

Router Filters

In addition to WEP, you can use filters that your router likely provides to allow only the network adapters you own to access your network.

Each network adapter you have, both wireless and wired, has a unique number burned onto the hardware's chip. You can configure your router to only let your adapters on the network by entering this number, called a MAC (media access control) address, into the software that comes with your router.

The MAC address on wireless network adapters is typically printed on the network adapter's underside (Figure 12.5).

FIGURE 12.5   Flip your network card over and you should see a clearly marked MAC address.

There are other non-network-specific ways to secure your data as well. We'll look at software you can install on your computers to make the entire wireless network safer.

How to Encrypt Your E-mail and Other Data

Encrypting e-mail and files on your system can help keep them secure. You can use freely downloadable software called PGP (Pretty Good Privacy) to encrypt the files on your hard drive, or the e-mail messages you send. You can download PGP at www.pgpi.org (Figure 12.6).

Like WEP, PGP requires a key that you create to encrypt your data. Unlike WEP, PGP is a two-key system requiring a public key and a private key. You have a key that you make public by posting on a web site, for instance, or e-mailing to someone. You also have a private key, which you do not distribute.

The public key is used to encrypt messages for you. You use your private key so that you—and only you—can decrypt your messages.

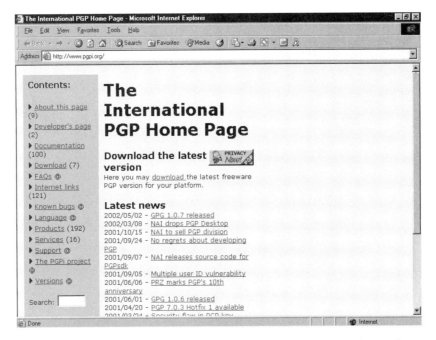

FIGURE 12.6 The PGPI web site offers the latest version of PGP for (free) download.

Viruses and Your Network

Viruses can spread quickly over a network, so your best defense is to install antivirus software on each computer on your network.

Vendors such as McAfee (www.mcafee.com) and Symantec (www.symantec.com) offer inexpensive (less than $40) antivirus software that can help keep your computers free of viruses.

Of course, you can do your part to keep your network clean of viruses:

- Be wary of attachments from people you don't know. If you don't know the person, don't open the attachment (Figure 12.7). Some viruses can start to open themselves, usually in the preview window of your e-mail program. You can turn off the preview feature of your mail program, or press the Esc key when asked if you want to open or save the file. Then *carefully* delete the attachment.

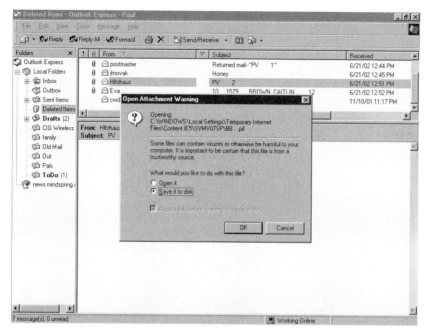

FIGURE 12.7 Don't open that e-mail! Hit the Esc key and immediately delete any message that looks suspicious.

- Today's viruses often come from someone you *do* know, who has accidentally opened a virus thinking it is a safe file. Make sure you know what the document is you're opening, even if it is from a trusted source.
- Worse, some viruses can "spoof" addresses, making it appear to have been sent from an address, when in fact it has been sent from a completely different address than the one you see. This is often done when the virus searches through an infected user's address book and picks an address at random to use as the sender's address. Check out Figure 12.7 and you'll see an e-mail that was created by a virus, and worse, has been spoofed to look like it's from someone I know. Tricky stuff.

Update your *definition list* often, your software's list of viruses that are currently in circulation. Active viruses are sometimes said to be "in the wild." Most antivirus software has automatic updating features that will connect over the Internet and download a current definition list (Figure 12.8). Make sure to use these features and get a new definition list often. Updating your list every 1–2 weeks will likely be sufficient.

If you find that a computer on your network is infected with a virus, check the Symantec and McAfee web sites for a downloadable software fix. The klez worm in particular is dangerous, because it has been widely distributed and can spread itself over a network. Shut down your Internet connection by turning off your modem, and get to work cleaning up your computers.

Common sense is your best defense. If everyone on your network knows to be careful with attachments, you should be looking good.

FIGURE 12.8 Use your antivirus software's features for updating definition lists of viruses that are in the wild.

Summary

Wireless networking by its nature requires more security than a wired networking technology, such as Ethernet. You can use some security methods, such as using a router and firewall to keep unauthorized users from gaining access to your network from the Internet. You can also use WEP to encrypt data over your wireless network. If you use a router, you can set up the router to only allow network adapters that you select to have access to your network. Finally, make sure each machine on your network has antivirus software installed, and be diligent in downloading up-to-date lists of circulating viruses, or your software will be of little use.

Appendix

Adding a Server

In this chapter...

✔ Wirelessly Connecting a New Server

✔ Setting Up File and Printer Sharing

✔ Adding a Wireless Printer Server

✔ Protecting Your Server

One of the greatest things about a home or small office wireless network is the ease in which you can set them up to communicate directly between computers. Unlike the networks used by many medium- to large-sized offices, your requirements probably don't necessitate using a server. However, if you have an extra computer, or would like to share data between home or office users without necessarily tapping your computer's resources each time someone connects, you can add a server to your network as easily as adding a wireless network adapter to a PC.

Many office networks use a powerful server to host software that is shared by many users (instead of installing the software on each individual PC). This makes the network administrator's job easier, and has other financial and ease-of-setup benefits. If you share data and printers among those connected to your network, you too might consider using a separate computer to handle data sharing and printing, centrally located in your wireless network.

Wirelessly Connecting a New Server

Using an old or new computer to share files and printers can make your wireless network quite a bit more efficient. Everyone in your home or office can access often-used files and printers without slowing any one person's computer on the network.

Adding the server is a simple as connecting a wireless network adapter to a PC.

Once your wireless network adapter is connected and working, you'll first need to name your computer. Then you can set up file and printer sharing. We discuss these steps in depth in Chapters 7 and 8. In the next sections, we'll look at the basic steps for getting your server up and running.

Setting Up File and Printer Sharing

Here we'll look at a quick rundown of the basic steps for sharing your server's files and printers. Windows 2000 and Windows XP offer print and file sharing turned on by default (if, for some reason, these features are not enabled, see Chapter 8).

Windows Me users can use the Home Networking Wizard. Choose Start, Programs, Accessories, Communications. Select the Home Networking Wizard. Windows XP offers its own Network Wizard (Start, All Programs, Accessories, Communications, Network Setup Wizard), which can step you through all the settings necessary for sharing your server.

To start sharing a Windows 98/98SE computer, follow these steps. Windows 95 users should also be able to get going using these steps.

1. From your Desktop, right-click Network Neighborhood and choose Properties (Figure A.1).

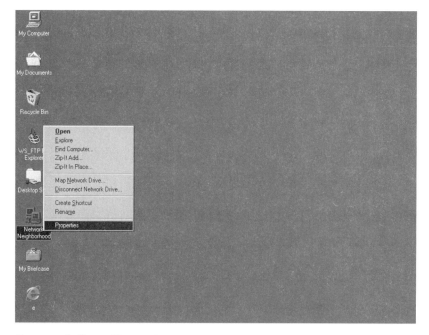

FIGURE A.1   Choose Properties to open the Network dialog box.

2. When the Network dialog box opens, click the Identification tab (Figure A.2).

3. Type in a computer and workgroup name. If you like, add a description that will let users know what type of computer is being shared, in this case, something like "The Server (Figure A.3)."

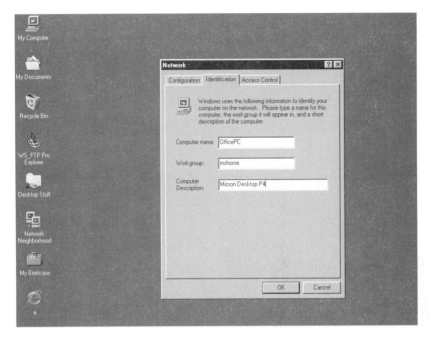

FIGURE A.2   Select the Identification tab.

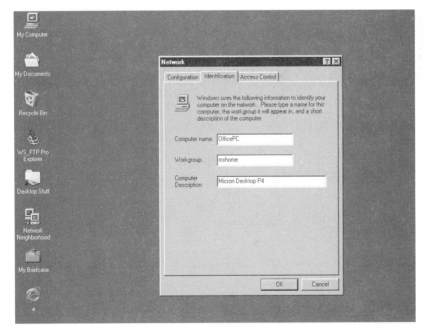

FIGURE A.3   Type in descriptive names to identify the server to users on your wireless network.

4. Click OK.

Now, let's turn on File and Print Sharing:

1. Right-click Network Neighborhood and select Properties (Figure A.4).

2. When the Network dialog box opens, click the File and Print Sharing button (Figure A.5).

3. Select "I want to be able to give others access to my files" and "I want to be able to allow others to print to my printer(s)." (Figure A.6)

4. Finish up by clicking the OK button (twice) to close the two open dialog boxes.

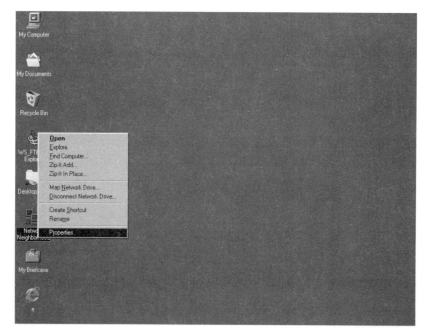

FIGURE A.4   Open Network properties.

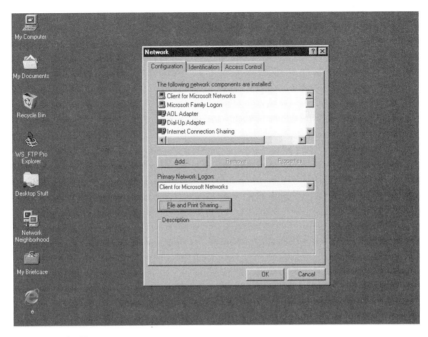

FIGURE A.5   Click File and Print Sharing.

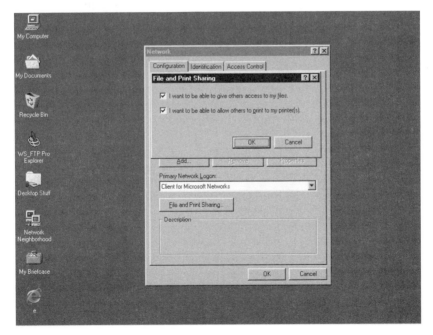

FIGURE A.6   Share files and printers.

Let's say that a printer is connected to the server for everyone to share. Let's go ahead and share it:

1. Open the Control Panel and select Printers (or Printers and Faxes in Windows XP) (Figure A.7).

2. Right-click the printer you want to share and choose Sharing (Figure A.8).

3. Select "Shared As." In XP choose "Share this printer." (Figure A.9)

Now we should "install" the printer on a computer that will be using the shared printer.

1. Select the shared printer from Network Neighborhood (Figure A.10).

2. Click the File menu and choose Install (Figure A.11).

3. The Add Printer wizard appears. Follow the directions of the wizard to start installing the drivers and start using the printer. You can also print a test page if you like (Figure A.12).

FIGURE A.7   Open the Printers control panel.

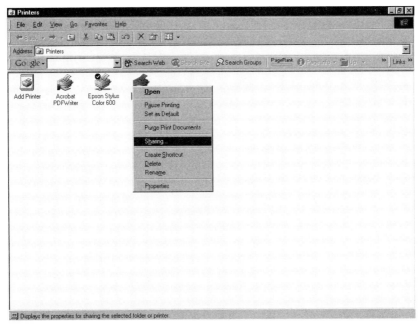

FIGURE A.8   Choose the printer you want to share by right-clicking it.

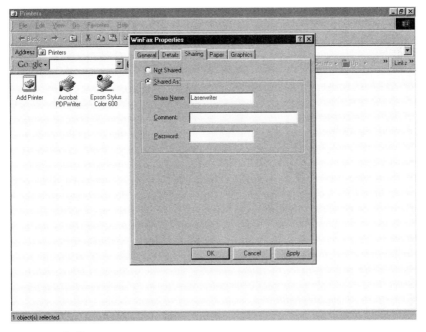

FIGURE A.9   Share it!

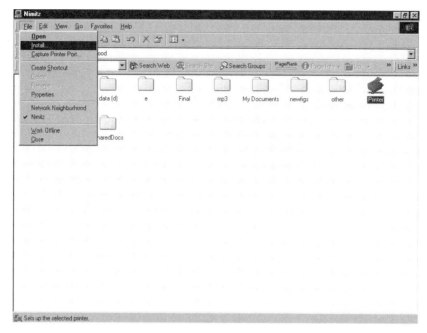

FIGURE A.10   Choose the shared printer.

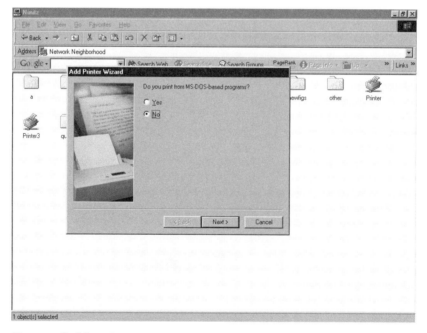

FIGURE A.11   Choose Install.

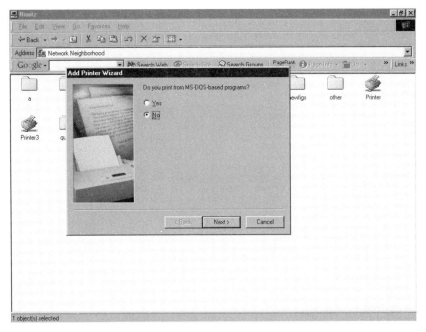

FIGURE A.12 The Add Printer wizard will guide you.

Start File Sharing on a Mac

Want to share files on your Macintosh? Here's how to enable file sharing in OS9:

1. From the Apple menu, choose Control Panels and select File Sharing.
2. From the Start/Stop tab in the File Sharing dialog box, enter your Owner Name and Owner Password.
3. Under File Sharing, click the Start button. Close the dialog box.
4. Open a folder to share, and then choose File, Get Info, Sharing.
5. Select "Share this item and its contents."

Adding a Wireless Printer Server

One of the coolest new wireless devices to hit the market is the wireless print server (Figure A.13). These devices are inexpensive (often less than $80) and provide a nice way to save time and space. Most home or small-office LANs connect one or more printers to a PC, then share that PC. It might be a PC that someone uses for work everyday, or a computer that is set aside for printing jobs only. Either of those situations can benefit from a wireless printer server. Here's why:

- You can place your printer server (and your printers) anywhere in your wireless LAN's range.
- You can free up a computer that you were using as a printer server.
- You can send multiple print jobs to the printer server, and it can handle the printing simultaneously.

FIGURE A.13 This wireless printer server from Linksys costs about $75. Netgear makes one for about the same price.

FIGURE A.14 You can also purchase a wireless access point (and router) with a built-in printer server. These bundled features are more and more common.

Adding a printer server can be a great way to share a printer, without having to leave one of your computers on to do it.

Protecting Your Server

Protecting your server is important if you share sensitive documents, such as financial records. Make sure to install a personal firewall (see Chapter 12). It's also a good idea to use a hardware firewall, such as the one built into many wireless access points/routers.

Summary

Wireless networking makes sharing a server easy. You can use an older, less-often-used machine as a server to share files and printers. In addition, wireless technology lets you place the server anywhere in your network's range.

Glossary

802.11a   A fast wireless networking technology, which transmits data at a maximum of 54 megabits per second. Also called Wi-Fi5. Wi-Fi5 is, as you might have guessed, about five times as fast as Wi-Fi equipment. 802.11a wireless networks use the less-crowded 5GHz band, and therefore are less prone to interference from home electronics.

802.11b   The most popular wireless networking standard, transferring data at up to 11 megabits per second. Also called Wi-Fi (Wireless Fidelity). This networking technology communicates over the unlicensed, 2.4-GHz radio band. 802.11b networks share the band with other home electronics including cordless phones and microwaves.

802.11g   A fast wireless networking standard that shares the 2.4GHz band with 802.11b equipment. 802.11g and 802.11b equipment can communicate with each other. 802.11g is faster than 802.11b (54 mbps rather than 11 mbps).

Access point   This hardware device allows wireless network cards to connect to a wired network. An access point has a wired component (an Ethernet port) and a wireless component (a radio that allows wireless network adapters to connect to the network).

Ad-hoc mode   Wireless networks typically work in one of two configurations (sometimes called *topologies*): ad-hoc or infrastructure. The topology or *mode* you choose depends on whether you want your PCs to communicate directly or with an access point. In ad-hoc mode data in the network is transferred directly to and from wireless network adapters connected to PCs. An ad-hoc network is also called a *peer-to-peer* network.

Bluetooth   A technology used to transfer data wirelessly. Bluetooth is found in PDAs and cell phones and can be used to transfer data to laptop and desktop computers and printers. Like 802.11b networking equipment, Bluetooth operates in the 2.4-GHz band, does not require a line of sight between components, and can pass through walls.

Broadband modem Unlike a 56-kilobit per second modem which sends and receives data over analog phone lines, a broadband modem sends a digital signal over your telephone or TV cable wiring, depending on whether you use a DSL or cable modem, respectively.

Cell Each wireless network device you use creates an area in which data is transmitted and received. These cells interlock providing greater distance over which the data is transferred, also called the equipment's *range*.

Data packet Data is transmitted over networks in pieces, called data packets. The data packet contains the data being sent as well as the address of the sender and recipient.

Ethernet A wired standard for networking hardware. Some of your equipment, such as a broadband modem, will connect to your access point by an Ethernet cable. Until recently, Ethernet was pretty much the only technology available for networking your computers.

Fast Ethernet A wired technology that can transfer data over cables at up to 100 megabits per second.

Firewall A firewall on your network puts a layer of protection between you and a hacker. Firewalls can be software you install on a computer, or they can be built into a router, or used as a standalone firewall hardware device. Whether you decide to install software or hardware, a firewall will help keep outsiders from accessing your network.

Gateway Hardware or software that allows multiple computers to access the Internet. In most cases, on a LAN, your gateway is a router. Your gateway could also be a single computer sharing its Internet connection with the other computers on the network.

Hub This hardware device is used to connect multiple wired elements of your network, including computers or routers, to your network. A hub has ports, usually four or more, into which you plug network cables. An access point works a bit like a wireless hub, and connects a wireless network to a wired network.

Hybrid network A network that mixes more than one networking technology. For example, you could create a network that uses Ethernet (wired) network adapters and 802.11b (wireless) network adapters.

Infrastructure mode You can increase the range of your wireless network by adding an access point. Wireless networks that use an access point are operating in infrastructure mode rather than ad-hoc mode, where the network adapters communicate directly with each other.

Internet connection sharing (ICS) A Windows feature bundled with operating systems Windows 98SE and later. The utility allows one of your computers to act as a router, sharing its Internet connection with other computers in your network.

IrDA   (Infrared Data Association)   A wireless networking technology that uses infrared light. Infrared is a line-of-sight technology, which, like your TV remote, requires devices to be lined up in a straight line to communicate.

LAN (Local Area Network)   A network of computers in one location, usually a home or office.

NAT (Network Address Translation)   Your router should have a feature called NAT, which allows you to share one IP address provided to you by your Internet Service Provider (ISP) for each computer on your network so that they can access the Internet.

Network adapter   Also called a network adapter card or Network Interface Card (NIC), this is a card installed in your computer that is used to connect the computer to a network.

MAC (Media Access Control) address   Each network adapter has its own unique serial number, called a MAC address. You can see the MAC address of your wireless network adapter—it's usually printed on the underside of the adapter.

PC Card   A removable module used to add features to a laptop, such as a network adapter, memory, or a small hard drive. A PC card slides into one of two PC Card slots you'll find on a laptop. Note that the PC Card slot is sometimes called a PCM-CIA slot (PCMCIA stands for Personal Computer Memory Card International Association).

PCI (Peripheral Component Interconnect) Card   A card you install into a slot in a desktop computer. PCI Cards are sometimes used to connect wireless network adapters to desktop computers. Connecting a PCI Card to a computer is slightly trickier than using USB. You must open the computer case and install the card in an open PCI slot inside the computer.

Protocol   A language used by a network to send and receive data. TCP/IP (Transmission Control Protocol/Internet Protocol) is the protocol used to transfer data over the Internet. You can also use TCP/IP as the protocol for your home network, for sharing Internet access, files, and printers.

Router   A hardware device or a software program that allows one network to connect to another. In a home network you can use a router to connect your LAN to the large network of interconnecting networks called the Internet. You can buy an access point with a built-in router. Your router will allow you to share a single Internet connection among all the computers connected to your network.

SSID (Service Set Identifier)   A name that identifies your network. To access the network, the SSID on each computer has to be the same.

Standard   An agreed-upon specification for the design of computer software or hardware. 802.11b is a wireless networking standard.

TCP/IP (Transmission Control Protocol/Internet Protocol)   The common language, or protocol, spoken by all computers on the Internet. On your home network, TCP/IP can be used both to access the Internet and to transfer files and share printers.

USB (Universal Serial Bus)   PCs that came with Windows 98 or later (when you purchased them) will have typically two USB ports. For connecting desktops to a 802.11b network, USB is a very good choice. USB network adapters are inexpensive and easy to install. The ports are hot-swappable, meaning you can plug in equipment and unplug it without rebooting the machine. That said, you will sometimes need to restart your computer for some network software to recognize the network adapter.

WAN (Wide Area Network)   A very large network spread over a large area, such as a cell phone network. When we talk about wireless networks we're talking about a home network that you can construct yourself, not a cellular voice or data network, which is often called a wireless WAN.

Wireless encryption   Networking hardware comes with software to encrypt data over the network so that it can't be read by an unintended recipient. The data is scrambled at the source, and then descrambled by the recipient. The technology standard for wireless encryption is called WEP (Wired Equivalent Privacy). You can enable wireless encryption on your network adapters and your access point.

Wireless networking   Connecting two or more computers to create a local area network (LAN), using radio transmitter/receivers (sometimes called transceivers).

Wi-Fi   See 802.11b.

Wi-Fi5   See 802.11a.

Photo Credits

Grateful acknowledgment is given to the following companies and organizations for the use of images of their products and Web pages in this book.

Chapter 1: p. 3 Linksys; **p. 6** Boingo.com; **p. 8** wi-fi.com; **p. 9** Proxim; **p. 10** Linksys

Chapter 2: p. 17 (top and bottom) Netgear; **p. 18** Netgear; **p. 19 (right)** Netgear; **p. 20** Linksys

Chapter 3: p. 33 D-Link; **p. 36** Linksys; **p. 37 (bottom)** Netgear

Chapter 4: p. 42 (top) 3Com; **p. 42 (bottom)** Netgear; **p. 44** 3Com; **p. 48** 3Com; **p. 49** Linksys; **p. 50** Linksys; **p. 51** Netgear

Chapter 5: p. 62 D-Link; **p. 64** Proxim; **p. 65** Netgear; **p. 67** Extended Systems

Chapter 6: p. 73 HomeNetHelp/Anomaly, Inc.; **p. 74** About.com; **p. 75** CNET; **p. 76** TechWeb; **p. 78** Shopper.com; **p. 79** PriceWatch; **p. 82** D-Link; **p. 83** Netgear; **p. 84** Google; **p. 85** paml.com; **p. 86** 802.11 Planet/Jupitermedia

Chapter 8: p. 109 Netgear

Chapter 9: p. 127 (top) Linksys; **p. 127 (bottom)** Netgear

Chapter 10: p. 148 (top) Yahoo!

Chapter 12: p. 173 SMC Networks; **p. 178** PGP International

Appendix: p. 192 Netgear; **p. 193** Linksys

Index

informIT

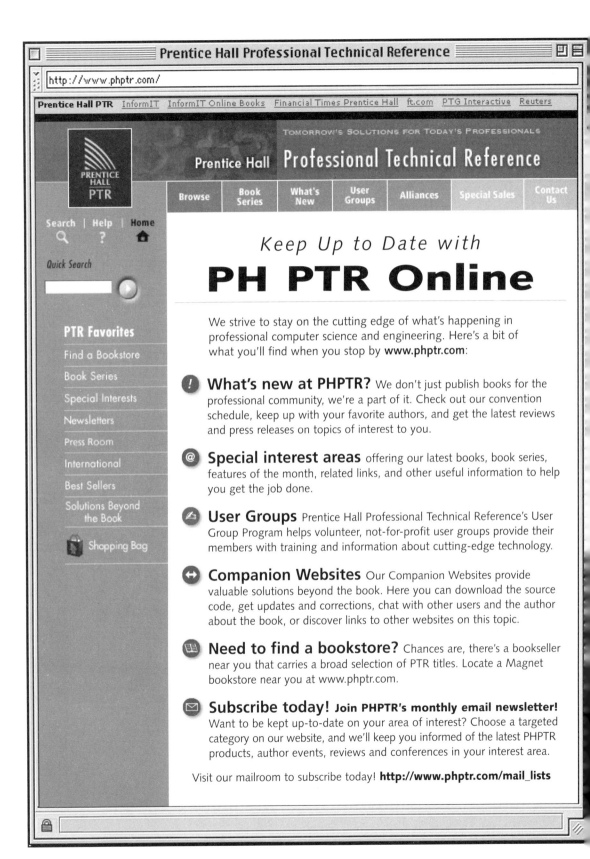